Mathematical Engineering

Series editors

Claus Hillermeier, Neubiberg, Germany
Jörg Schröder, Essen, Germany
Bernhard Weigand, Stuttgart, Germany

For further volumes:
http://www.springer.com/series/8445

Michael Robinson

Topological Signal Processing

Michael Robinson
Department of Mathematics and Statistics
American University
Washington, DC
USA

ISSN 2192-4732 ISSN 2192-4740 (electronic)
ISBN 978-3-662-52284-4 ISBN 978-3-642-36104-3 (eBook)
DOI 10.1007/978-3-642-36104-3
Springer Heidelberg New York Dordrecht London

Softcover reprint of the hardcover 1st edition 2014

Printed on acid-free paper

Springer is part of Springer Science+Business Media (www.springer.com)

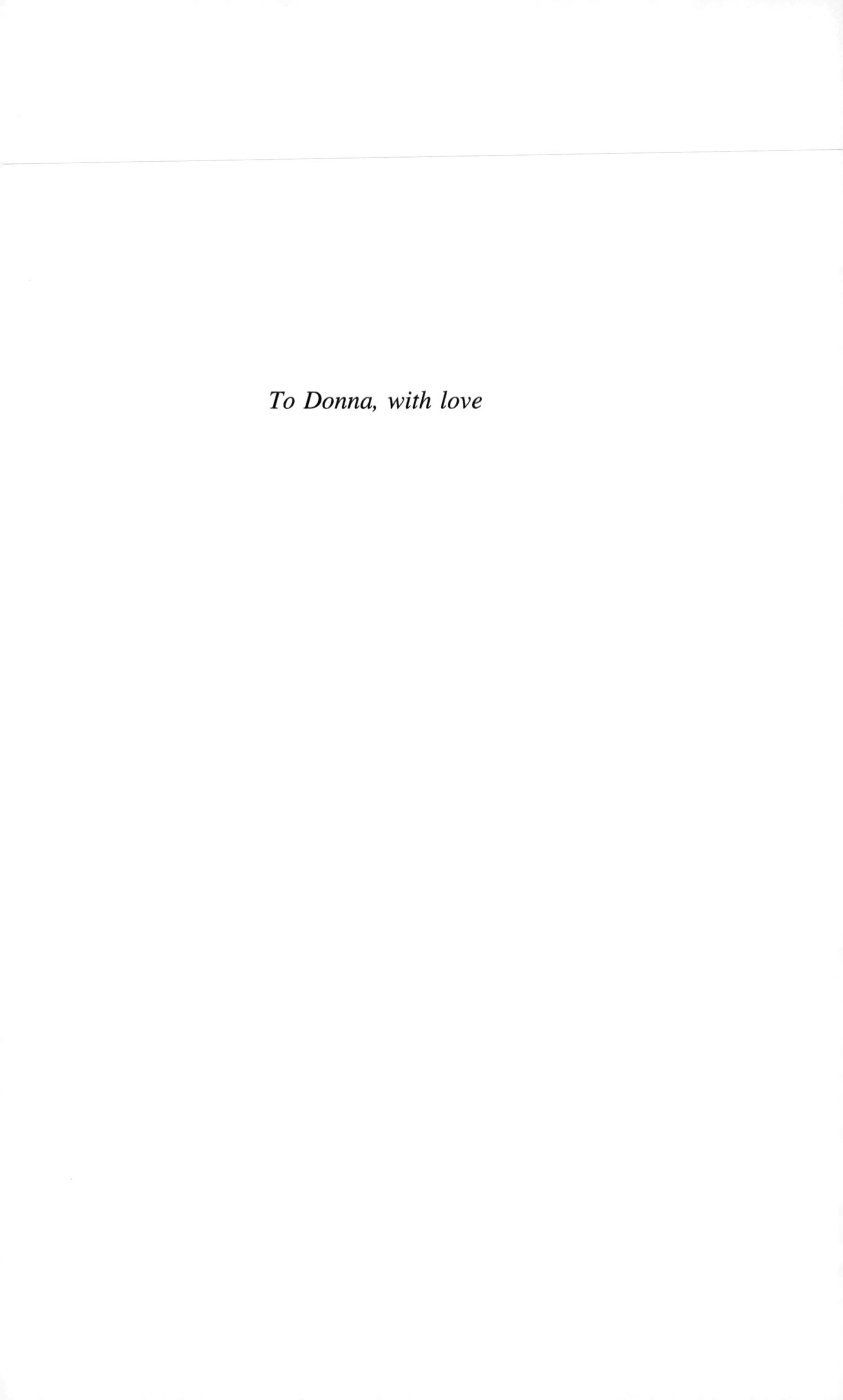

To Donna, with love

Preface

"Topological" is an unlikely adjective to be associated with signal processing, which is dominated by results flowing from geometric rigidity. However, heterogeneous sensors and sensors in heterogeneous environments present a unique challenge; there is too much uncertainty for rigidity to be of much use. This book takes the perspective that signal processing has much to gain by taking a more *local* approach; consistency between nearby sensors or measurements is expected, but is not expected between sensors that are far apart. But how does one measure *distance* without explicitly invoking geometry, which is potentially very uncertain? This is the purview of *topology*; the lesson is that nearness can be studied implicitly and local signals can be studied through the theory of *sheaves*.

Sheaves have an unduly bad reputation, even among mathematicians, so it was unexpected that they could unify a number of signal processing concepts. Therefore, this book places sheaves at the conceptual and computational center by focusing exclusively on sheaves over cellular spaces. In this context (which is somewhat more limiting than the usual definition), many of the venerable results about sheaves still hold, but the proofs are substantially easier. Focusing on sheaves means that the exposition treats *cohomology* almost exclusively, which is more natural than singular or cellular homology for signals. For these reasons, it is usually easy to connect a local signal-processing concept to the correct sheaf theoretic tool.

Because the author is both a practitioner as well as a theoretician, due consideration is given to implementation concerns. Each chapter contains at least one *case study* in which the theoretical tools are used to address a relevant engineering problem. In most cases, the case study outlines a prototype implementation developed by the author that uses simulated or experimental data.

This book is intended for first-year graduate students and advanced undergraduates in mathematics and engineering. As such, a background including both linear algebra and multivariable calculus is expected and is used to motivate the necessary topological concepts. Algebraic topology relies on point-set topology, at least theoretically. There is an appendix on point-set topology, though intuition is usually a good guide to the necessary concepts. In a few places, a passing familiarity with abstract algebra and the structure of the ring $\mathbb{Z}$ is helpful but never necessary, since another appendix reviews the necessary background.

Because the book brings together many dissimilar concepts, it is meant to be read in a linear fashion. Those readers comfortable with the material can skim ahead, but should be wary. Of necessity, the treatment differs from the traditional one both from a mathematical and an engineering perspective. That said, proofs, remarks, and exercises are called out carefully so that they may be omitted on first reading. No essential details are "buried" inside proofs, and some of the more technical or less insightful proofs are merely referenced or sketched. Because of this format, each topic is introduced informally, defined precisely, and then explained through examples. Reading the examples is essential for mastery of the subject. Engineering is built through a combination of theory and practice; the examples highlight both.

Finally, I welcome comments, suggestions, and corrections from you, the reader. Feel free to send me a message! I maintain a list of emendations on my website http://www.drmichaelrobinson.net/ which you may find useful.

Washington, DC, December 2013 Michael Robinson

Acknowledgments

I am grateful to Rob Ghrist for suggesting that I write this book, and for reassuring me that writing it would not be an impossible task. The experience of writing has been both clarifying and inspiring. My wonderful wife Donna really made this book possible by her persistent encouragement.

I would like to thank my mentors, both in academia and in industry. A few of them are worthy of special mention: John Hubbard, my thesis adviser, from whom I first learned about sheaves and cohomology; Rob Ghrist, my postdoctoral adviser and fellow applied topologist; and finally my friends and colleagues at SRC, Inc.: D. J. Isereau, Mark Perillo, Sean O'Hara, and Eva Piltch-Boucher, whose continued collaboration with me has assured me that sophisticated mathematics has a place in the real world.

I am thankful to the many people who read portions of this book and offered extremely helpful suggestions as it was being written, especially Rob Ghrist, Donna Dietz, and Morgan DeHart.

Finally, I gratefully acknowledge the financial support for this book and the work that led to it, which includes internal funding from SRC, Inc. and the following Federal contracts HR0011-09-1-0050, HR0011-07-1-0002, FA9550-09-1-0643, N000140810668.

Contents

Symbols

$(M, \mathscr{U})$	A manifold M with an atlas $\mathscr{U}$
$(V_\bullet, f_\bullet)$	A sequence of vector spaces
$*$	The point added to compactify a space, or an external vertex
$/$	Quotient of two spaces
1_A	The indicator function on a set A
$< v, w >$	An inner product
argmax	The location of the first maximal value in a list
$\rightsquigarrow$	An attachment between two cells
$\setminus$	Set difference
χ	The Euler characteristic
$\circ$	Composition of two functions
coker	The cokernel of a matrix
deg	Degree of a vertex
Δ^k	The standard k-simplex
δ_x	Dirac measure concentrated at x
dep	The depth of a cover
dim	Dimension of a manifold
ed	Edge distance between two vertices in a graph
$\mathscr{F}$	The face category of a cell complex
id	The identity map
image	The image of a matrix or a function
inf	Infimum of a set, its greatest lower bound
$\mapsto$	Defines the input and output of a function
$\mathbb{C}$	The complex numbers
$\mathbb{N}$	The natural numbers (including 0)
$\mathbb{RP}^k$	k-dimensional real projective space
$\mathbb{R}$	The real line
$\mathbb{Z}$	The integers
med	Maximal edge distance between two vertices in a graph
$\oplus$	Direct sum of spaces, or sum of sheaves
$\otimes$	Tensor product of spaces or sheaves

$\bar{A}$	The closure of a subset A
∂	The boundary of a topological subspace or manifold
$\perp$	Value which indicates failure to receive a signal
pr	A projection map
rank	The rank of a matrix
reach K	The reach of a subset K of $\mathbb{R}^n$
$\mathscr{PL}$	The sheaf of piecewise linear functions on a cell complex
$\mathscr{PS}^k$	The k-th persistence sheaf
$\mathscr{S}$	Script type is used for sheaves or collections of sets
$\mathscr{S}(U)$	The sections of a sheaf $\mathscr{S}$ on a set U
$\mathscr{S}^Y$	The sheaf constructed from sampling stalks of $\mathscr{S}$ on a subcomplex Y
$\mathscr{S}_Y$	The ambiguity sheaf for the morphism $\mathscr{S} \to \mathscr{S}^Y$
$\sim$	An equivalence relation
$\sqcup$	Disjoint union
star	The star over a cell in a cell complex
sup	Supremum of a set, its least upper bound
supp	The set of support for a function
span	The span of a collection of vectors
$C((a,b))$	The continuous, real-valued functions on the open interval (a,b)
$C^k(X;\mathscr{S})$	k-cochains of a sheaf $\mathscr{S}$ on X
$C_c^k(X;\mathscr{S})$	Compactly supported k-cochains of a sheaf $\mathscr{S}$ on X
$CF(X,Y)$	The space of constructible functions from X to Y
D^k	The k-dimensional closed unit disk
$f^*\mathscr{S}$	The pullback of a sheaf $\mathscr{S}$
$f_*\mathscr{S}$	The pushforward of a sheaf $\mathscr{S}$
$H^k(V_\bullet)$	The k-th cohomology of a sequence of vector spaces
$H^k(X;\mathscr{S})$	The k-th cohomology of a sheaf $\mathscr{S}$ over X
$H_c^k(X;\mathscr{S})$	The compactly supported k-th cohomology of a sheaf $\mathscr{S}$over X
$H_k(V_\bullet)$	The k-th homology of a sequence of vector spaces
K^ε	The ε-offset of a set K
$N(\mathscr{U},R)$	The nerve of $\mathscr{U}$ witnessed by a set R
S^k	The k-dimensional unit sphere
X_f	A cell complex on whose cells a function f is constant
Cat	Boldface type is used for categories
EFh, **EB**h	Boldface is used for Euler integral transforms of a function h
Mor(**C**)	Morphisms of a category **C**
Obj(**C**)	Objects of a category **C**
$\boldsymbol{R_\varepsilon P}$	The Vietoris–Rips complex of size ε

Chapter 1
Introduction and Informal Discussion

Signal processing is the discipline of extracting information from related collections of measurements. To be effective, measurements must be organized and then filtered, detected, or transformed in some way to expose the desired information. Distortions from uncertainty and noise oppose these techniques, and degrade the performance of practical signal processing systems. Statistical methods have been developed to ensure consistent performance of these systems, but usually need large amounts of data in aggressively uncertain situations.

What if systems could be developed that are completely insensitive to distortion? We should not expect to obtain complete information in these settings, but can useful questions still be answered? It happens that they can, by using topological tools—those that model spaces only up to continuous transformations. Since the collection of continuous transformations is large and varied, we should expect that tools which are topologically-motivated should be automatically insensitive to substantial distortion. A change in perspective is needed, since topology is both *qualitative* and *precise*. Although the topological mindset sparks the discovery of powerful algorithmic ways to treat data, it may appear foreign and surprising at first to those accustomed to more traditional signal processing.

This book has three major goals:

1. To show that topological invariants provide qualitative information about signals that is both relevant and practical for system analysis and design.
2. To show that the signal processing concepts of filtering, detection, and noise correspond to the seemingly abstract mathematical concepts of sheaves, functoriality, and sequences, respectively.
3. To advocate for the theory of sheaves, bringing it to the forefront of signal processing theory and practice.

M. Robinson, *Topological Signal Processing*,
Mathematical Engineering, DOI: 10.1007/978-3-642-36104-3_1,

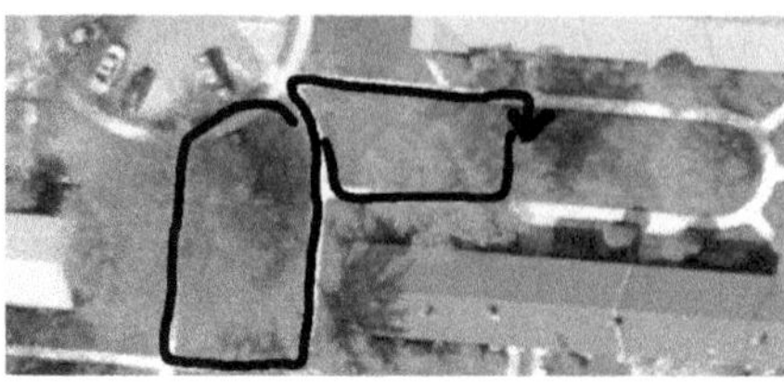
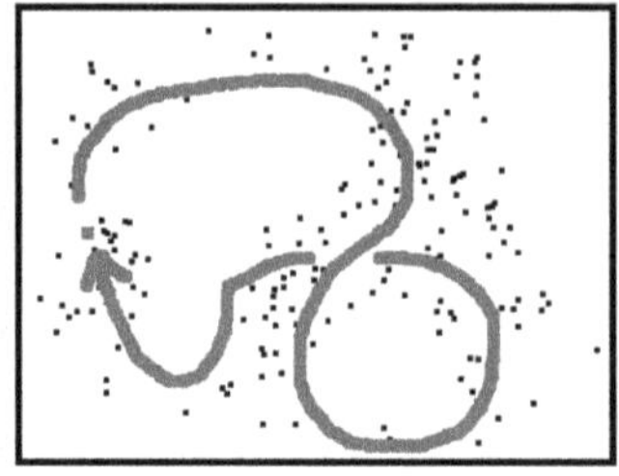

Fig. 1.1 A path on which the author walked while taking signal strength measurements of nearby wireless access points (*left*) and a 2-dimensional projection of the data, in which his path's topology is visible (*right*)

1.1 Meet the Case Studies

Although this book contains both theorems and proofs, the focus is practical. The direct application of theoretical tools to practical systems draws their benefits and shortcomings into sharp focus, and highlights where new tools must be developed. Therefore, each chapter is built around several case studies that emphasize the practical aspects of the techniques developed in that chapter. The applications center around the author's own work in remote sensing, localization, and sonar imaging.

1. *Signal manifolds for localization, tracking, and navigation* (Sect. 2.3). This case study examines the use of ambient radio signals to localize a receiver. Each location of the radio receiver and orientation of its antenna yields a different collection of signal strengths for each transmitter that is visible. If enough independent measurements are taken, the resulting information uniquely determines the location of the receiver and orientation of its antenna. This idea is quite powerful, and can be used to detect if a mobile receiver has returned to a given location as shown in Fig. 1.1.
2. *Topological filters* (Sect. 3.4). The "processing" of signal processing is usually performed by filters, which are often linear and translation invariant. We show how to generalize these filters using sheaves; in doing so, we obtain a new class of nonlinear, angle-sensitive image filters.
3. *Indoor acoustics and quantum graphs* (Sect. 3.5). This case study shows how to organize the signal strengths associated to wave propagation in a convenient way. This organizational improvement permits the extraction of distances between transmitters and receivers, and distances along propagation paths. We use this mathematical formulation to explain an algorithm that processes narrowband signals, and scales linearly with the number of edges in the graph.
4. *Tracking water pollution* (Sect. 4.6). The Shannon-Nyquist theorem can be interpreted as a consequence of the sheaf model of signals. However, the sheaf theoretic model of sampling additionally applies to sampling on networks. We demonstrate this model by showing how to track water pollutants to their source.

5. *Extracting topology from intersections in coverage* (Sect. 4.7). This case study builds on Case Studies 1 and 3, and adds the capability to construct a topologically accurate, combinatorial model of the space of locations for the receiver. By using knowledge of which transmitters can be decoded simultaneously, topological detectors and filters help to construct this model with relatively little effort.
6. *Target enumeration* (Sect. 5.2). Knowing the number of vehicles that pass a collection of sensors is an important part of traffic engineering. Sensors called "traffic counters" tabulate the number of vehicles that cross over a rubber hose anchored on the road, without recording any identifying information specific to the vehicle. If the counter registers the correct number of counts when multiple vehicles cross the hose simultaneously, then it is possible to accurately recover the total number of vehicles that cross over a longer period of time. This idea can be generalized to larger, dense networks of sensors that produce counts of the objects in their immediate vicinity. This case study shows how the *Euler characteristic integral* provides a theoretically justified (and easy to implement) way to obtain the total number of detected objects from such a sensor network.
7. *Shape recognition in computer vision* (Sect. 5.4) can be treated topologically. Traditional matched filter methods can detect congruent copies of a desired feature in an image, but they fail if the feature's size or orientation is changed. Euler integral transforms can be developed that are insensitive to size or rotation; only being sensitive to a feature's shape. By tailoring (or "tuning") a particular Euler integral transform, it is possible to adjust its selectivity for particular shapes in an image.
8. *Experimental validation of topology extraction* (Sect. 6.2). This case study addresses the problem of noise tolerance in Case Studies 1, 3, and 5. Sadly, the algorithm for topological reconstruction (Case Study 5) is not very noise tolerant. We show how to improve it by separating significant topological features from those that are likely to be spurious, by using *persistent cohomology*. This case study presents an experimental study that validates the robustness of persistent cohomology. Figure 1.2 shows the acoustic laboratory in which data for this case study was collected.
9. *Quasiperiodic signals* (Sect. 6.4). This case study shows how rotating targets can be detected even if their motion is not periodic, which occurs when the rotation rate varies. Traditional methods for detecting rotation in this case tend to fail. This case study uses acoustic echos from a ceiling fan collected by a cell phone to demonstrate performance of a topological algorithm that robustly detects rotating targets.
10. *Recovery of a space from measurements of waves* (Sect. 6.6). This case study ties together the previous ones, by showing the modes of failure and correct operation of topological algorithms in the presence of noise. Specifically, when measurements of acoustic and radio signal strength are used to recover the topology of an environment, topological errors can be caused by undersampling.

The sheer diversity of examples where topological signal processing makes a contribution is remarkable. While it is true that many of these examples could

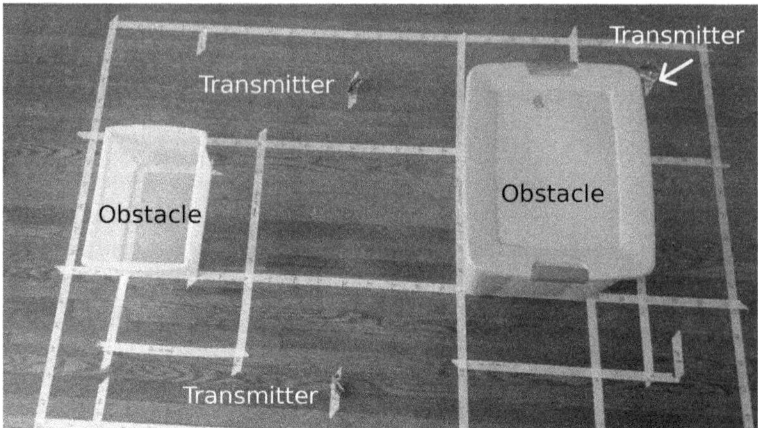

Fig. 1.2 Setup for an acoustic obstacle detecting experiment, used in Case Study 8. There are three transmitters and two obstacles. The receiver is not shown

be approached in an *ad hoc* manner without invoking topology, the power of a topological approach is that it *unifies* the common themes among them. This common framework provides a fertile ground for cross-disciplinary collaboration on widely varying practical problems.

1.2 Discrete Approximations of Spaces

As a mathematical theory, the study of topology is based on the examination of *topological spaces*, which specify a consistent notion of the *neighborhoods* of a point. In signal processing, topological spaces describe the sensors and their utilization during the measurement process. If the phenomenon being measured is compatible with the topological space, then sensors within a given neighborhood will measure consistent readings in the absence of noise. This provides a solid context for describing filters that reject noise by averaging.

The physical world in which signal processing systems operate imposes many strong constraints on the underlying topological spaces of sensors. This book primarily focuses on *cell complexes*, which are simply a collection of closed disks of various dimensions glued together along their boundaries. Several examples are shown in Fig. 1.3. We will usually find that an "exploded view" of a space called the *the attachment diagram* (see Fig. 1.4) is a key tool for manipulating cell complexes.

Cell complexes offer a good balance between flexibility and the avoidance of pathology. All of the complexity of a cell complex lies in its attachment diagram, since it is assembled from convenient, topologically simple cells. We will usually find that that this combinatorial structure is more than sufficient to represent any of the spaces we desire. The payoff is a powerful theory for manipulating signals with

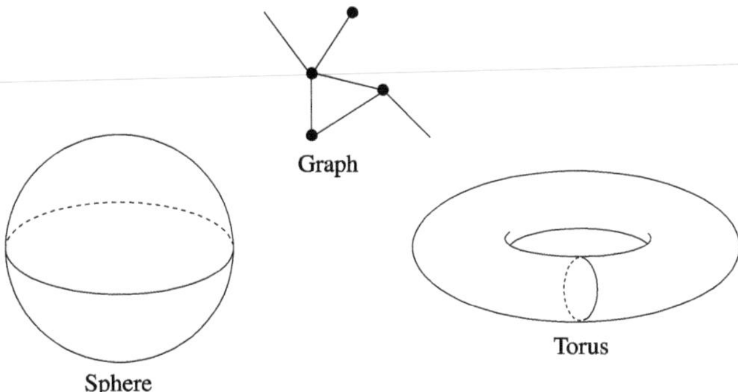

Fig. 1.3 Example of several cell complexes

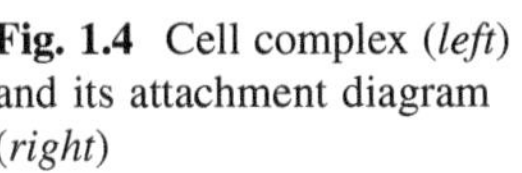
Fig. 1.4 Cell complex (*left*) and its attachment diagram (*right*)

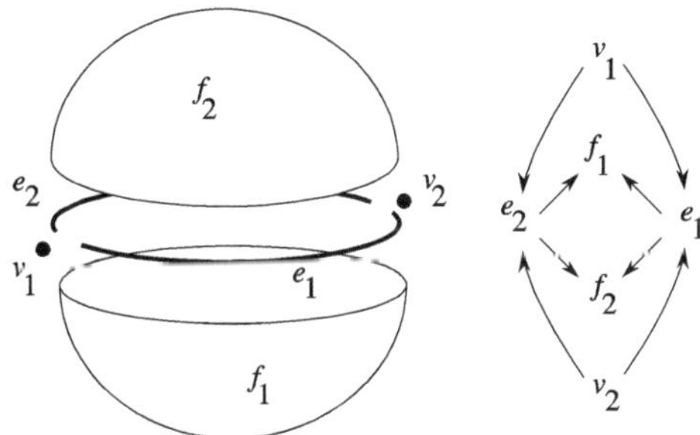

minimal outlay; many celebrated and useful results are much easier to understand in this context.

Sometimes it is useful to work with *manifolds*, a class of spaces that are even better behaved than cell complexes. Manifolds are general spaces which have enough structure to support the familiar operations of calculus. Because physical problems are often described by differential equations, manifolds play an important supporting role in joining more traditional concepts of signal processing to the topology. (We do not lose much generality by this approach, since a manifold can be given the structure of a cell complex.)

1.3 Local Data: Filtering

A signal consists of a collection of related measurements. Usually, not all of the measurements in a signal are closely related; a given measurement is typically only related to its neighbors in time, space, or frequency. However, most measurements (even those of remote phenomena) are taken locally. For instance, the temperature and pressure at a particular location are local, as is the electric field strength of a propagating wave. Therefore, the appropriate model and subsequent operations on signals should respect the local structure of measurements.

It makes sense to use a mathematical construct that respects locality but is theoretically lightweight. The appropriate structure is that of a *sheaf*. Although often relegated to obscurity (even within mathematics), a sheaf is merely a way to assign different kinds of data to each part of a space and to check consistency of this data between neighboring locations. In essence, sheaves are the correct data structure in which to store local data and the correct abstraction of a *topological signal model*.

The signals themselves are rarely the most important feature of a signal processing system. The filters, detectors, and transformations that make up the system are more important. Signal processing systems can be represented as sequences of sheaf-based *topological filters*, which generalize the concept of discrete, linear, translation-invariant systems. Indeed, topological filters describe every manipulation of signals (not just timeseries or images) that preserves locality, and are the basis of several new and powerful algorithmic approaches to filter design.

1.4 The Interplay Between Local Data and Global Inference: Detection

Much of the data in any signal is not of interest in a given problem; it is clutter. The effectiveness of sheaves in signal processing hinges on their ability to reject or mitigate clutter. For instance, consider the familiar situation of trying to listen to a friend in a busy room. Usually, there is plenty of sound energy from your friend that reaches your ears. The problem is the presence of many other conversations that you also hear. This unwanted, but structured, portion of a signal is called *clutter*. Practical usage of a signal processing system requires the removal of this clutter.

The solution to the clutter problem usually relies on the internal consistency of the desired signal. The unique characteristics of your friend's voice and patterns of speech make it possible for your brain to extract it from many other competing voices. Sheaves are particularly good at describing internally consistent information. Their precise definition allows locally consistent signals to be assembled into globally consistent ones, automatically removing inconsistent clutter.

This leads to a compelling definition of *detectors*; those operations that assemble local data consistently into global data. Detectors summarize and extract the information from a signal, leaving a highly compressed representation that preserves the relationships between signals. When correctly designed, this representation reduces the amount of information in the signal by rejecting clutter.

This text advocates for sheaf *cohomological invariants* as being the most prominent class of detectors. By using *sequences* of topological filters (signal *processing chains* instead of merely signals), robust cohomological invariants arise. Applying detectors to these processing chains yields the notion of *persistent cohomology*, which is a compelling way to detect topological features in datasets.

This book also connects these cohomological invariants to signal reconstruction problems. For instance, the celebrated Shannon-Nyquist theorem guarantees that a signal can be reconstructed from its samples if it is bandlimited. Similarly, compressive sensing approaches explain how certain signal models support sparser sampling. Both of these ideas can be expressed by stating that local measurements containing complete information about a signal are reduced to those that lie on a lower-dimensional subspace, which can be represented as a sequence of topological filters. Studying the cohomology—a detector operating on these filters—leads to a general condition for perfect reconstruction of a signal from samples on that lower-dimensional subspace.

A different way to develop robust filters is to use the time-honored concept of averaging. For instance, the Euler characteristic χ of a simplicial complex can be used to build a consistent theory of integration. Once an integral has been defined, integral transforms are natural and interesting tools for signal processing. Using the Euler characteristic integral leads to non-local ways to manipulate local data that allow the combined use of geometric constraints and topological flexibility.

1.5 Coda: An Invitation

Each chapter ends with a list of open questions, since statements of goals and problems often lead to unexpected solutions. The lists are incomplete and many of the questions are ill-specified. The author hopes these open questions will spark ideas, discussion, and progress in the new and developing field of topological signal processing.

Chapter 2
Parametrization

This chapter develops several models of topological spaces associated to collections of measurements that

1. Parametrize the context in which the measurements are taken,
2. Set the stage for the understanding when measurements are consistent, and
3. Are convenient to implement in applications.

Local signals are collections of measurements which are related by their spatial, temporal, or contextual proximity. This leads one to study the sets in which "proximity" has a consistent mathematical meaning, namely *topological spaces*. The general definition of a topological space, while elegant and intellectually efficient, sadly admits many pathologies. (Indeed, it admits so many pathologies that at least one book, Steen and Seebach (1978), has been written about them!) Because of this, we will study several restricted classes of topological spaces which admit fewer pathologies. Of course, there could be a situation of interest which is not admitted in these classes, but the author has not yet found an example of such a situation. Additionally, these classes of spaces are well-suited to algorithmic manipulation, and therefore play an important role throughout this book.

The primary class of topological spaces that are used in applications are *cell complexes*. These include some *CW complexes* and all *simplicial complexes* as shown in Fig. 2.1. From a computational perspective, simplicial complexes are quite convenient to manipulate, though in high-dimensional settings they can incur a large memory footprint. It is for this reason that others, for instance Kaczynski et al. (2004), also consider cubical complexes. However, most of the cases of interest for signal processing use lower dimensions, so we will avoid the additional technical overhead of discussing cubical complexes.

When more structure is available in the problem, it is sometimes useful to instead study *manifolds*, which are spaces that are locally like Euclidean space. Conveniently, every manifold can be given the structure of a cell complex, and every compact manifold can be given the structure of a CW complex. Unlike cell complexes and CW complexes, manifolds admit a general notion of differential calculus. Because manifolds admit both a differential and a combinatorial interpretation, manifolds

M. Robinson, *Topological Signal Processing*,
Mathematical Engineering, DOI: 10.1007/978-3-642-36104-3_2,

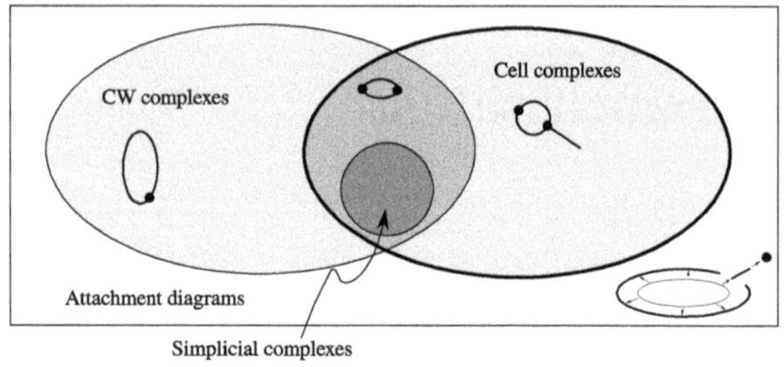

Fig. 2.1 Some kinds of topological spaces that are important in applications

can be useful in cementing the relationship between a physical signal model (arising from differential equations) and the algorithmic, combinatorial models we develop in this book.

2.1 Abstract Spaces

We begin our study of restricted topological spaces with *CW complexes*. These spaces are assembled from disks glued along their boundaries in a particular way. In the end, we will focus on related spaces, called *cell complexes*, because these are the most natural place to study sheaves and local signals.

2.1.1 CW Complexes

In order to connect computations concerning the structure of a space to data, it is essential to have a constructive, combinatorial description of a space that arises from attaching simpler topological spaces together.

Definition 2.1 Suppose that X and Y are topological spaces and that $f : A \subseteq X \to Y$ is a continuous function from a subspace of X. Let $Z = (X \backslash A) \sqcup Y$, and define the function $g : X \sqcup Y \to Z$ by

$$g(p) = \begin{cases} p & \text{if } p \in Y \text{ or } p \notin A \\ f(p) & \text{if } p \in A \end{cases}$$

The *adjunction space* is the topological space $(Z, \mathscr{U})$, whose topology $\mathscr{U}$ consists of all subsets of Z whose preimages are open in $X \sqcup Y$. We say that Z consists of X and Y *attached* along $A \sim f(A)$. (See Fig. 2.2 for a schematic of this construction.)

General attachments can be rather complicated, so we will instead consider spaces that arise from attaching disks to one another along their boundaries.

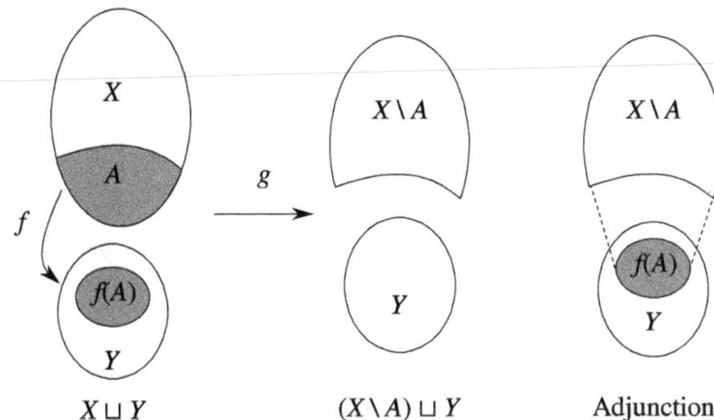

Fig. 2.2 An adjunction of two spaces

Definition 2.2 The n-dimensional *closed disk* is the closed subset $D^n = \{(x_1, \dots, x_n) \in \mathbb{R}^n : x_1^2 + \dots + x_n^2 \leq 1\}$. The boundary of such a disk is the $n-1$-dimensional *sphere* $\partial D^n = S^{n-1}$, the closed subset $\{(x_1, \dots, x_n) \in \mathbb{R}^n : x_1^2 + \dots + x_n^2 = 1\}$. (We will use the notation ∂A to represent the boundary of a set A.) Similarly, the n-dimensional *open disk* is the subset $\{(x_1, \dots, x_n) \in \mathbb{R}^n : x_1^2 + \dots + x_n^2 < 1\}$.

Definition 2.3 Suppose that the topological space X^k consists of the (not necessarily disjoint) union of a collection $\mathcal{K}$ of disks of dimension at most k, and that ∂D^{k+1} is decomposed as a finite union $\partial D^{k+1} = A_1 \cup A_2 \cdots \cup A_n$. An *attaching map*[1] or *attachment* is a continuous function $\phi : \partial D^{k+1} \to X^k$, for which the image $\phi(A_i)$ of each A_i is a disk, and each such image is in the collection $\mathcal{K}$.

We then say that D^{k+1} is *attached* to X^k and to each of the disks $\phi(A_i)$, and will write $\phi(A_i) \rightsquigarrow D^{k+1}$. Observe that the symbol $\rightsquigarrow$ goes in the opposite direction as the attaching map itself, a convention that is useful when we study sheaves.

The concept of a *CW complex* captures the idea of inductively constructing a space from lower dimensions to higher dimensions using attaching maps, as suggested in Fig. 2.3.

Definition 2.4 Suppose that X^0 is a disjoint collection of points and that X^k is the adjunction space formed by attaching finitely many disks to X^{k-1} for $k = 1, \dots, n$ as above. Then $X = X^n$ is a *finite CW complex* of dimension n, or simply a *CW complex*. The space X^k is called the *k-skeleton* of X.

Attachment constructions like these are powerful tools for understanding how a space is assembled from its parts. They play a similar role to exploded view diagrams (see Fig. 2.4) in assemblies. Because of this graphical connection, attachment constructions can be visualized with an *attachment diagram*, showing how the cells are attached, such as in Fig. 2.5. Each cell is shown and the links represent attachments (drawn from low dimensional cells to higher dimensional cells). Usually, we suppress any attachments that are compositions of other attachments.

[1] The term *map* will be used as shorthand for "continuous function" throughout the book.

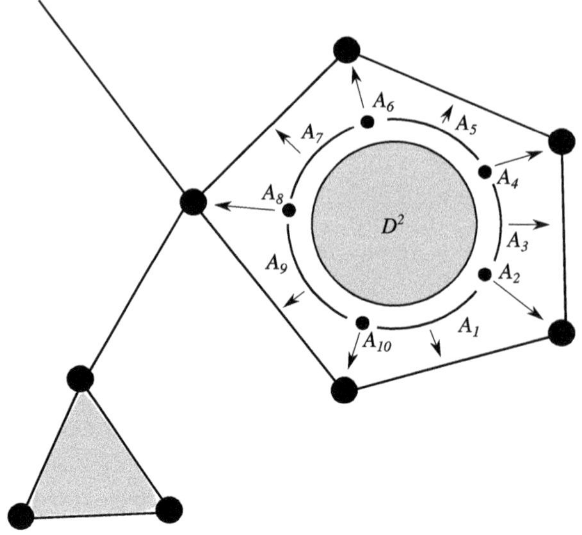

Fig. 2.3 Example of an attaching map (marked with *arrows*) of a disk D^2

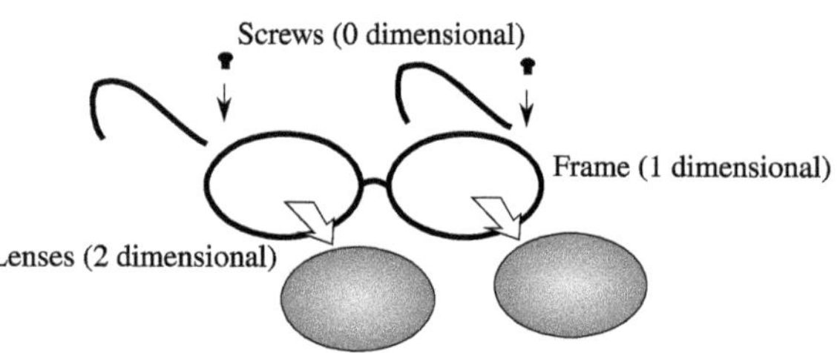

Fig. 2.4 An attachment construction is like an exploded view for a topological space

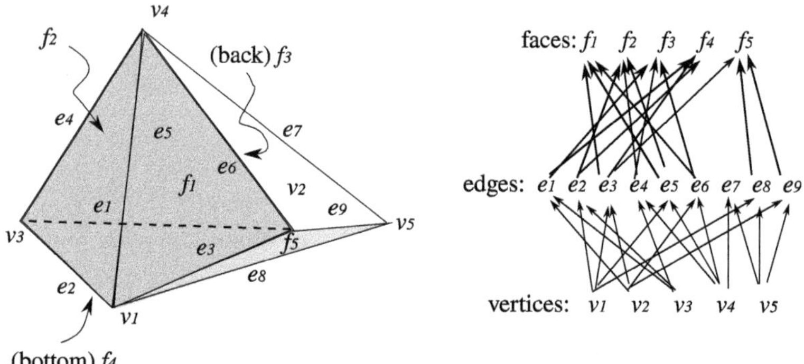

Fig. 2.5 An attachment construction (*left*) and its attachment diagram (*right*)

Remark 2.1 The attachment diagram of an attachment construction is a set that is partially ordered by the sequence of attachments. When we exhibit an attachment diagrams, we display its Hasse diagram.

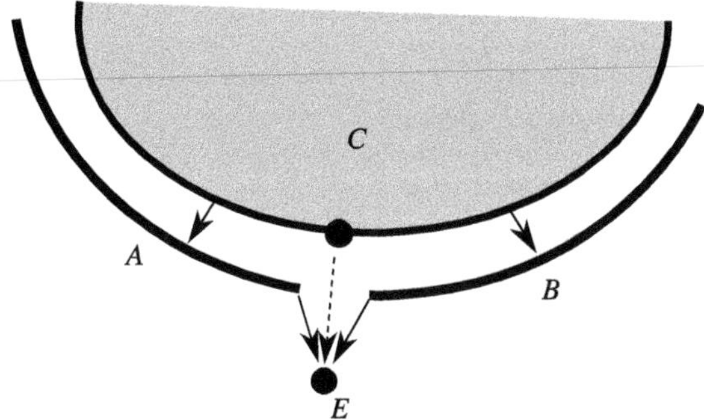

Fig. 2.6 The compatibility condition for attaching maps in CW complexes; the attaching map from the boundary of C to A also restricts to the attachment of A to E

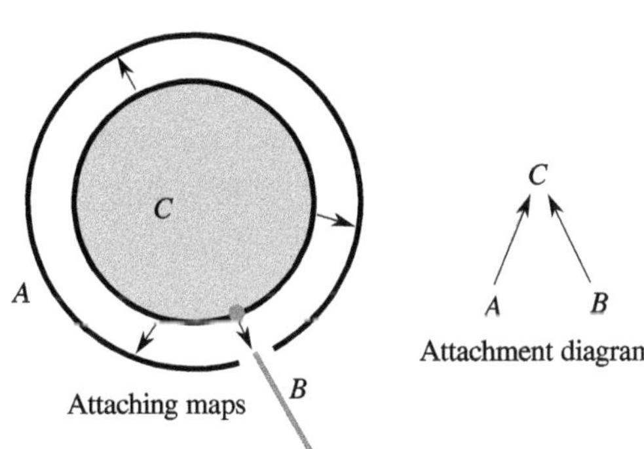

Fig. 2.7 An attachment construction that is not a CW complex. Note that B is an edge attached to a 2-dimensional disk C without a vertex between them

Because we assume that attaching maps are continuous, no new attachments between cells in X^k arise when we attach higher dimensional cells. As shown in Fig. 2.6, if two cells A and B are both attached to a higher dimensional cell C at the same place on C, then they must also be attached to the same *lower* dimensional cell E there also. If we remove the higher dimensional cell C, the attachment diagram retains the common attachments of A and B to E.

Remark 2.2 If the attaching maps are not assumed to be continuous, the resulting construction would not necessarily be an adjunction space. For example, the disks A and B in Fig. 2.7 would become detached from one another if the higher-dimensional cell C were to be removed. With the cell C present, the 1-skeleton of the attachment construction is a different topological space than with C removed.

Corollary 2.1 *The k-skeleton of a CW complex is also a CW complex.*

The boundaries of closed disks are CW complexes, and they can be easily assembled inductively. To start the induction, we begin by constructing a circle as a CW complex. This structure consists of two 1-cells (closed intervals) $e_1 = [0, 1]$ and $e_2 = [0, 1]$, attached to two 0-cells (points) v_1 and v_2 as shown in Fig. 2.8. Specifically, there are two attaching maps, namely

- $a_1 : \partial e_1 = \{0, 1\} \to X^0 = \{v_1, v_2\}$ given by $a_1(0) = v_1$ and $a_1(1) = v_2$, and

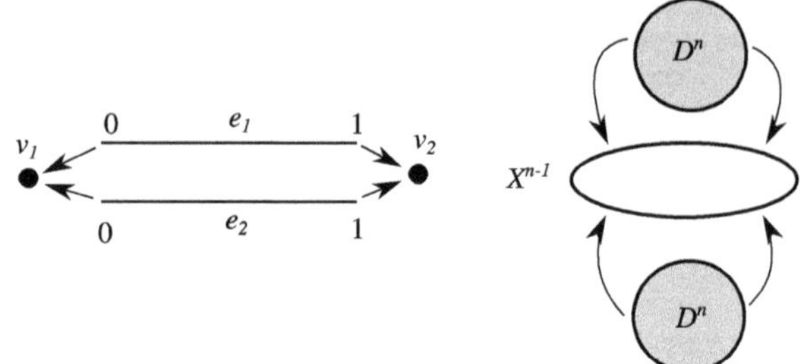

Fig. 2.8 CW complex structures for a circle (*left*) and sphere (*right*). (From a geometric perspective, this construction of S^2 looks more like a "whoopie cushion" than a sphere.)

- $a_2 : \partial e_2 = \{0, 1\} \to X^0 = \{v_1, v_2\}$ given by $a_2(0) = v_1$ and $a_2(1) = v_2$.

To construct an n-dimensional sphere, we assume that the $n-1$-skeleton X^{n-1} consists of a single $n-1$ dimensional sphere as a CW complex. We then attach two n-dimensional disks by attaching the boundary of each to X^{n-1} with a homeomorphism as shown in Fig. 2.8.

Other constructions are possible as well. For instance, the n-dimensional *torus* is constructed by attaching opposite faces of the n-dimensional cube to each other. This is not quite as constrained of a definition as it might first seem; how precisely the attachment is made does matter. Consider the two-dimensional situation, in which opposite sides of a square are attached to each other. If, for instance, $(0, x)$ is attached to $(1, x)$ and $(x, 0)$ is attached to $(x, 1)$, the resulting space is a torus. However, if instead we consider the CW complex that arises from attaching $(0, x)$ to $(1, x)$ and $(x, 0)$ to $(1-x, 1)$, the resulting space is called a *Klein bottle* and is quite different. The Klein bottle is a space which cannot be homeomorphically mapped into a 2-dimensional subset of $\mathbb{R}^3$. A standard projection of its embedding[2] into $\mathbb{R}^4$ to $\mathbb{R}^3$ is shown in Fig. 2.9.

Exercise 2.1 The *real projective n-plane* $\mathbb{RP}^n$ arises as a quotient of the closed n-disk D^n by the equivalence $v \sim -v$ for each $v \in \partial D^n$. Construct a CW complex structure for $\mathbb{RP}^n$.

While flexible, CW complexes have the disadvantage that they are sometimes too constrained. This is especially the case when studying graphs; a CW complex representation of a graph requires both endpoints of every edge to be connected to a vertex. Sometimes, however, it is more descriptive to remove a vertex from the end of an edge. This might signal that the edge's connection on that end is not useful to consider, or might indicate a connection outside of the region being considered. The appropriate generalization is that of a *cell complex*. A cell complex is a either (1) a CW complex whose attaching maps are embeddings or (2) a space that becomes a CW complex after being compactified, and whose attaching maps are embeddings even once the space has been compactified.

[2] An *embedding* is an injective continuous map $f : X \to Y$ whose image $f(X)$ is homeomorphic to its domain X.

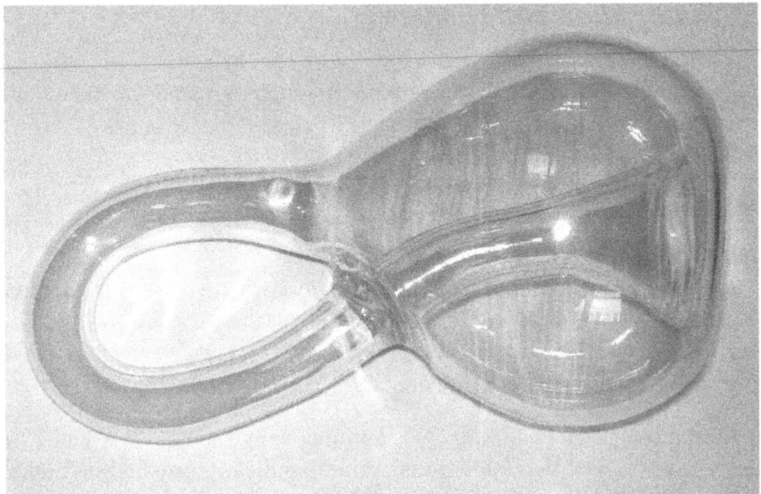

Fig. 2.9 Projection of an embedded Klein bottle into $\mathbb{R}^3$

Fig. 2.10 A cell complex (*left*) that is not a CW complex. Its one-point compactification (*right*) has the structure of a CW complex

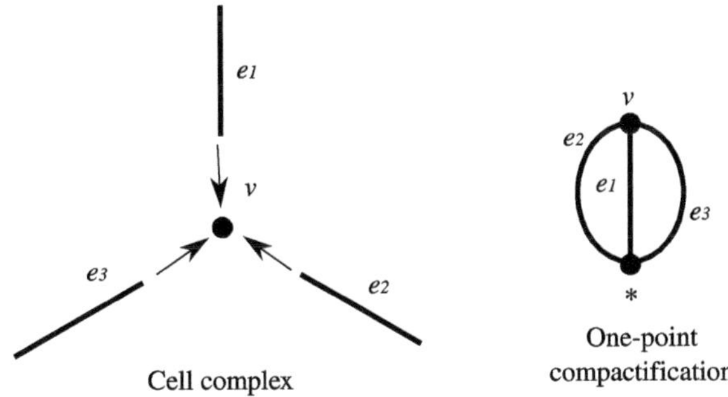

Definition 2.5 A *cell complex* consists of a topological space X and a collection of disjoint *open* disks $\{c_k\}$, each of which is a subspace of X, such that if we add an additional *external vertex* $*$ and form $\{\overline{c_k}\} \cup \{*\}$, these are the cells of a CW complex structure for the one-point compactification $X \cup \{*\}$ (see Proposition A.7 in the Appendix) of X, whose attaching maps are all embeddings. We call the open disks $\{c_k\}$ *open cells* whenever there is the possibility of confusion.

Figure 2.10 shows a cell complex that is not a CW complex, since only one endpoint of each edge is attached to a vertex. However, its one-point compactification is a CW complex.

The simpler space, consisting of a single open interval is not a cell complex, however. Its one-point compactification is a circle, but its only attaching map is not an embedding since both endpoints are attached to the same (new) point. This problem can be rectified by subdividing the open interval into two open intervals attached to a common point in the middle.

2.1.2 Cellular Maps and Homotopy

Cell complexes can be related to one another in much the same way that continuous maps relate topological spaces.

Definition 2.6 A continuous map $f : X \to Y$ between two cell complexes is called a *cellular map* if

1. For each cell $c \in X$, $f(c)$ is a cell in Y,
2. If $a \rightsquigarrow b$ is an attachment of two cells in X then $f(a) \rightsquigarrow f(b)$ is an attachment in Y, and
3. Given a cell $c \in X$ and two other cells $d, e \in f(c)$, then $f^{-1}(d) \cap \overline{c}$ is compact if any only if $f^{-1}(e) \cap \overline{c}$ is also compact.

It is worth noting that condition (3) is automatically satisfied if X and Y are CW complexes, and the last two conditions are automatically satisfied if X and Y are both abstract simplicial complexes (Sect. 2.2.1).

Topology is popularly understood to be the branch of mathematics that deals with deformations, particularly of images like the sequence shown in Fig. 2.11. This is an example of a *homotopy* from one image to another.

Definition 2.7 A *homotopy* between two maps $f : X \to Y$ and $g : X \to Y$ is a continuous function $H : X \times [0, 1] \to Y$ in which $H(x, 0) = f(x)$ and $H(x, 1) = g(x)$ for all $x \in X$. We call f and g *homotopic* (written $f \simeq g$) if there is a homotopy between f and g.

The fact that a single topological space may be described by several different cell complex structures raises the question of how to tell if two cell complexes are the "same". One can ask whether two spaces are homeomorphic (or stronger, homeomorphic through a cellular map), but this is too stringent of a criterion. CW complexes are often best studied under a different criterion, called *homotopy equivalence*.

Definition 2.8 A *homotopy equivalence* between two topological spaces X and Y is a pair of maps $f : X \to Y$ and $g : Y \to X$ such that both $f \circ g$ and $g \circ f$ are homotopic to identity maps. We say that X and Y are *homotopy equivalent* if there exists a homotopy equivalence between them.

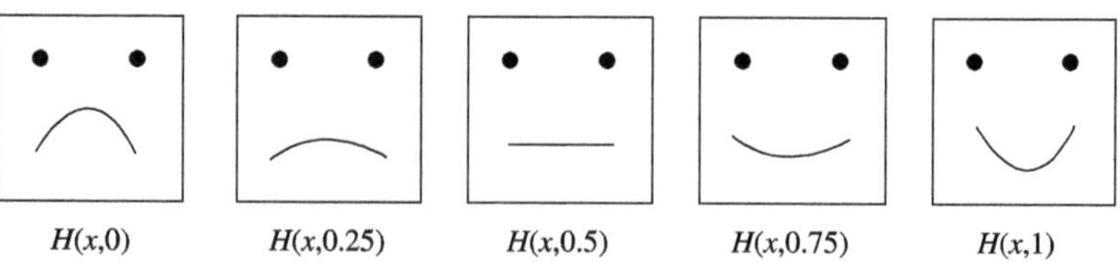

Fig. 2.11 A continuous deformation of one image into another

Exercise 2.2 Show that $[0, 1]$ is homotopy equivalent to itself, by showing that the compositions of $f, g : [0, 1] \to [0, 1]$ given by

$$f(x) = \sqrt{3x} \text{ and } g(y) = 4y^2$$

are homotopic to the identity map $\mathrm{id}_{[0,1]}(x) = x$.

Rather than directly checking if the resulting topological spaces associated to an attachment construction are homotopic, it is usually easier to look for algebraic "homotopy invariants" that allow different spaces to be discriminated. If two spaces are homotopy equivalent, they have the same sets of invariants; so if the invariants associated to two spaces are different, then the spaces cannot be homotopy equivalent. However, the converse is not true; if two spaces have the same set of invariants, they may still not be homotopy equivalent.

For CW complexes, the invariants which discriminate between the most spaces are found in homotopy groups. Unfortunately, computing homotopy groups is computationally and theoretically very difficult, and there remain many open questions about how to effectively perform the computation at all. On the other hand, much of the complexity can be avoided by using weaker invariants called *homology* and *cohomology*. While easier to compute, homological invariants cannot discriminate between certain homotopy *in*equivalent spaces. Nevertheless, we will develop and use homology theory throughout the rest of the book because of its computational utility.

2.2 Representation of Spaces

As convenient as CW complexes are for constructing and manipulating spaces, they are remarkably difficult to represent completely in software. Because of this, it is useful to have more restricted descriptions of spaces that have the same notion of topology as CW complexes, but are more computationally convenient. From a theoretical point of view, one can switch between equivalent representations as appropriate.

2.2.1 Abstract Simplicial Complexes

The prototypical combinatorial model of a space is an *abstract simplicial complex*. It is extremely easy to store and manipulate an abstract simplicial complex in a computer because it can be represented using ordered lists.

Definition 2.9 An *abstract simplicial complex* X on a set A is a collection of ordered[3] lists of A that is closed under the operation of taking sublists. We call each element of X, which is an ordered collection of elements of A, a *simplex*. A simplex with $k+1$ elements is called a k-simplex or a k-face, and we call a 0-face a *vertex* and a 1-face an *edge*. A *simplicial map* $f : X \to Y$ between an abstract simplicial complex on A to an abstract simplicial complex on B is a function induced on simplices by a function $A \to B$.

Example 2.1 Suppose that $A = \{p, q, r, s, t\}$.

- Then $X = \{\{p, q\}, \{p\}, \{q\}, \{r\}\}$ is an abstract simplicial complex on A.
- Although $X' = \{\{q, p\}, \{p\}, \{q\}, \{r\}\}$ consists of the same *subsets* as X, we regard X and X' as being different simplicial complexes, since the ordering of the elements is different.
- $Y = \{\{p, q\}, \{p\}\}$ is not a simplicial complex, since it is missing the 0-simplex $\{q\}$.
- Both $Z = \{\{p, q\}, \{p, r\}, \{r, q\}, \{p\}, \{q\}, \{r\}\}$ and
- $W = \{\{p, q, r\}, \{p, q\}, \{p, r\}, \{r, q\}, \{p\}, \{q\}, \{r\}\}$ are abstract simplicial complexes.

Example 2.2 Suppose that $X = \{\{p, q\}, \{r, s\}, \{p\}, \{q\}, \{r\}, \{s\}\}$ and $Y = \{\{p, q\}, \{p\}, \{q\}\}$ are two simplicial complexes on $A = \{p, q, r, s\}$ and that $\phi : A \to A$ is a function given by

$$\phi(p) = p, \ \phi(q) = q, \ \phi(r) = p, \ \phi(s) = q.$$

This function induces the following simplicial map:

$$\{p, q\} \mapsto \{p, q\}, \ \{p\} \mapsto \{p\}, \ \{q\} \mapsto \{q\}, \ \{r, s\} \mapsto \{p, q\}, \ \{r\} \mapsto \{p\}, \ \{s\} \mapsto \{q\}.$$

While very efficient, the definition of an abstract simplicial complex sometimes obscures topological information. It is therefore important to be able to *realize* a finite abstract simplicial complex X as a CW complex $|X|$ in which simplicial maps become cellular maps.

As shown in Fig. 2.12; each simplex in X corresponds to a cell in $|X|$, and is attached to each of its faces. We begin by constructing the realization of a single k-simplex.

Definition 2.10 The *standard k-simplex* is the closed subset of $\mathbb{R}^{k+1}$ given by

$$\Delta^k = \left\{ (x_1, \dots x_{k+1}) : x_i \geq 0 \text{ for all } i, \text{ and } \sum_{i=1}^{k+1} x_i = 1 \right\}.$$

[3] The elements of A are need not be ordered themselves.

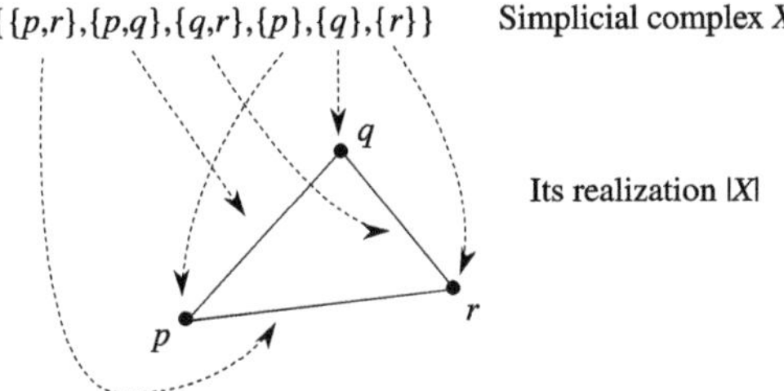

Fig. 2.12 A realization of an abstract simplicial complex

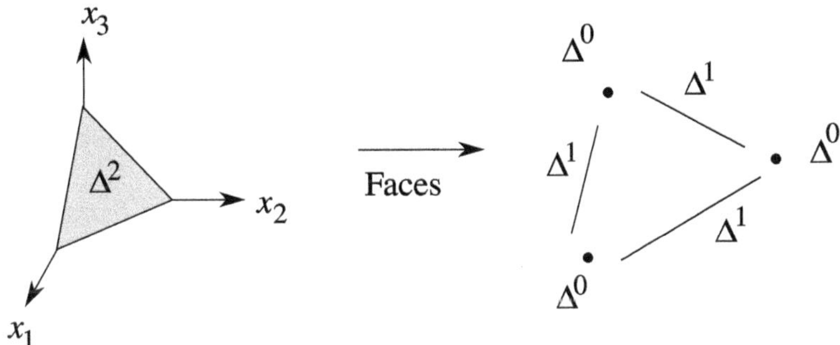

Fig. 2.13 The standard 2-simplex Δ^2 and its faces: Δ^0 are *points* and Δ^1 are *line* segments

Observe that Δ^k is the convex combination of the $k + 1$ points whose coordinates are all 0 except one which is 1.

As can be seen in Fig. 2.13, intersecting the standard 2-simplex with the cardinal planes $x = 0$, $y = 0$, or $z = 0$ results in copies of the standard 1-simplex. This is generally true, since

$$\Delta^{k-1} = \Delta^k \cap \{(x_1, \ldots, x_{k-1}, 0)\}.$$

By permuting coordinates in the intersecting subspace, a total of $k + 1$ copies of the standard $k - 1$-simplex form the boundary of the standard k-simplex. It follows that intersecting Δ^k with a subspace formed by zeroing out n coordinates results in a copy of Δ^{k-n}.

Definition 2.11 Let X be an abstract simplicial complex on the finite set A. We define the *realization* $|X|$ to be the attachment construction given by $|X| = \left(\bigcup_\alpha \sigma_\alpha\right) / \sim$, where α is an index that ranges over all simplices in X and $\sigma_\alpha = \Delta^{k_\alpha}$ is a standard simplex of dimension k_α. We construct the attachment maps $\phi_\alpha : \partial\sigma_\alpha \to |X|^{k_\alpha - 1}$ so that each $(k_\alpha - 1)$-face of σ_α is mapped homeomorphically onto an element of $|X|^{k_\alpha - 1}$ as follows. Without loss of generality, suppose that σ_α corresponds to the k-simplex $\{1, 2, \ldots, k\}$. Consider ϕ_α restricted to the face $\{1, 2, \ldots, k - 1\}$. Since X^k is an abstract simplicial complex, the simplex $\{1, 2, \ldots, k - 1\}$ is also contained in X^k, and therefore corresponds to some other simplex σ_β. We therefore assign

$$\phi(\partial\sigma_\alpha) = \phi_\alpha(x_1, x_2, \ldots, x_{k-1}, 0) = (x_{j_1}, x_{j_2}, \ldots, x_{j_{k-1}}) \in \Delta^{k-1} = \sigma_\beta, \quad (2.1)$$

where $(1, \ldots, k-1) \mapsto (j_1, \ldots, j_{k-1})$ is a permutation. We note that assembling each ϕ_α from its restrictions to faces of σ_α (given by (2.1)) is automatically continuous, and it is therefore an attaching map.

Notice that since abstract simplicial complexes are closed under the operation of taking subsets, the restriction of ϕ_α to the boundary of the preimage of σ_β is automatically an attaching map. Further, each such ϕ_α is an embedding by construction. Hence $|X|$ is a cell complex.

Proposition 2.1 *If X is a CW complex whose attaching maps are embeddings, then there is an abstract simplicial complex Y for which $X \to |Y|$ is cellular homeomorphic to $|Y|$.*

In other words, for any CW complex X, one can find an abstract simplicial complex whose realization is a topological copy of X.

The statement is true if the attaching maps are not embeddings, though the proof is more difficult. We will not need that level of generality, however.

Proof We can proceed by induction on attaching maps. A 0-dimensional complex can be realized as a collection of vertices. If X^k can be realized as $|Y^k|$, consider attaching a $k+1$-cell e_α to X^k via the map ϕ. Suppose that $A_1 \cup \cdots \cup A_n$ is a decomposition such that all open cells in $\phi(e_\alpha)$ are one of the $\phi(A_i)$. If $n \leq k+2$, it is immediate (since ϕ is an embedding) that we merely need to add a single $k+1$-simplex to Y whose faces correspond to the $\phi(A_i)$. If $n > k+2$, then we merely need to subdivide e_α into simplices so that the union has n faces. (Figure 2.14 shows the 2-dimensional case.) □

Corollary 2.2 *If X is a noncompact cell complex, then it is cellular homeomorphic to $|Y| \setminus \{*\}$, the geometric realization of some abstract simplicial complex Y with a vertex removed. We will call $*$ the* external vertex.

Corollary 2.3 *If $f : X \to Y$ is a simplicial map, then f induces a cellular map $|f| : |X| \to |Y|$.*

2.2.2 Manifolds and Embeddings

While cell complexes have an overall notion of dimension, their dimension is not local. For example, the interior of an edge in a graph is homeomorphic to an open interval; a 1-dimensional space. Perhaps at a vertex attached to one or two edges, the dimension ought to be 1. But what about a vertex that is attached to three edges? It is unclear what dimension should be assigned to such a vertex. This situation does not arise in some applications, which constrains the allowable local topological structure in those cases. This leads naturally to the concept of a manifold.

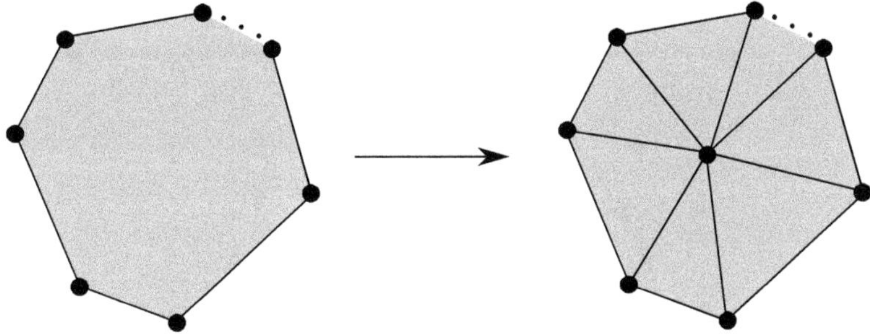

Fig. 2.14 Subdividing a 2-cell with n edges into simplices

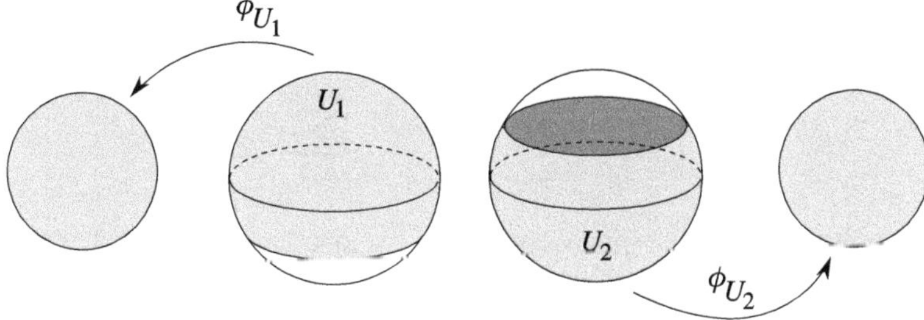

Fig. 2.15 An atlas with two charts on the 2-dimensional sphere S^2

Our study of CW complexes is intimately tied to the structure of manifolds; a compact manifold of a particular dimension can be given the structure of a CW complex of that dimension. (This is why we will assume manifolds to be Hausdorff[4] and paracompact,[5] since CW complexes satisfy both properties.) The most elegant proof of this fact involves the use of Morse theory. The author recommends several excellent texts on the subject, for instance Milnor (1963), Banyaga and Hurtubise (2004).

Definition 2.12 A *topological manifold* $(M, \mathscr{U})$ is a Hausdorff topological space M that is paracompact and has an *atlas*, a cover $\mathscr{U}$ of open sets in which each $U \in \mathscr{U}$ is homeomorphic to an open subset V of Euclidean space $\mathbb{R}^n$ through a map $\phi_U : U \to V$, called a *chart of* $\mathscr{U}$ (n may depend on U and is called the *local dimension* at U *and is written* $\dim U$).

Example 2.3 A sphere is an example of a manifold. The n-dimensional sphere is the following closed subset of $\mathbb{R}^{n+1}$

---

[4] A space is Hausdorff if every two distinct points are contained in disjoint open neighborhoods.

[5] A space is *paracompact* if every open cover $\mathscr{U}$ has a locally finite open refinement cover $\mathscr{V}$ in which each element $V \in \mathscr{V}$ is a subset of some $U \in \mathscr{U}$.

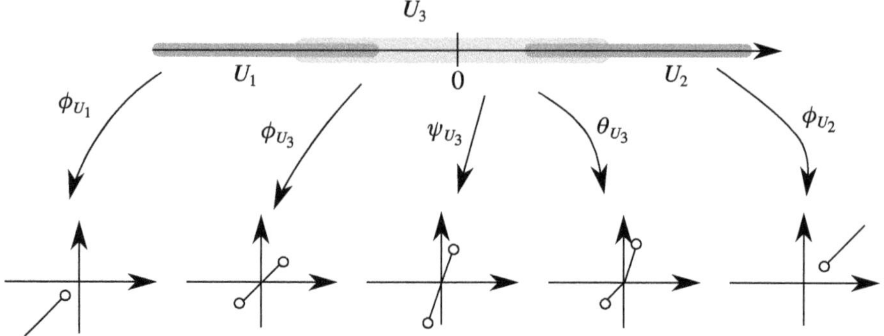

Fig. 2.16 Several different charts for $\mathbb{R}$

$$S^n = \left\{ (x_1, \ldots, x_{n+1}) \in \mathbb{R}^{n+1} : \sum_{i=1}^{n+1} x_i^2 = 1 \right\}.$$

One convenient atlas for the sphere consists of two overlapping hemispheres U_1, U_2, as shown in Fig. 2.15. (This atlas is analogous to the CW complex for the sphere shown in Fig. 2.8.) It's easiest to define ϕ_{U_1} and ϕ_{U_2} using the stereographic projection:

$$\phi_{U_1}(x, y, z) = \left(\frac{x}{1+z}, \frac{y}{1+z} \right) \text{ for } z > -1$$

and

$$\phi_{U_2}(x, y, z) = \left(\frac{x}{1-z}, \frac{y}{1-z} \right) \text{ for } z < 1.$$

Definition 2.13 Suppose that $\mathscr{U}$ is an atlas of a topological manifold M. The *transition maps* of $\mathscr{U}$ are the functions $\phi_{U_1} \circ \phi_{U_2}^{-1}$ for $U_1, U_2 \in \mathscr{U}$, which are maps between open subsets of Euclidean space. If the transition maps have k continuous derivatives, then we say that $\mathscr{U}$ is a C^k *atlas*. If the transition maps have derivatives of all orders, then we say that $\mathscr{U}$ is a C^∞ *atlas* or a *smooth atlas*.

Exercise 2.3 Show that the transition maps $\phi_{U_1} \circ \phi_{U_2}^{-1}$ and $\phi_{U_2} \circ \phi_{U_1}^{-1}$ in Example 2.3 , which are functions $\mathbb{R}^2 \backslash \{(0, 0)\} \to \mathbb{R}^2$, both have continuous derivatives away from the origin.

Definition 2.14 Two C^k atlases $\mathscr{U}$ and $\mathscr{V}$ are C^k *compatible* if their union $\mathscr{U} \cup \mathscr{V}$ is also a C^k atlas.

Example 2.4 This example shows several different atlases that can be put on the real line $\mathbb{R}$. Consider two open sets, $U_1 = (-\infty, -1)$, $U_2 = (1, \infty)$ with charts $\phi_{U_1} : U_1 \to \mathbb{R}$, $\phi_{U_2} : U_2 \to \mathbb{R}$ given by $\phi_{U_1}(x) = x$ and $\phi_{U_2}(x) = x$. We now consider another open set $U_3 = (-2, 2)$ with three different possible charts (see Fig. 2.16),

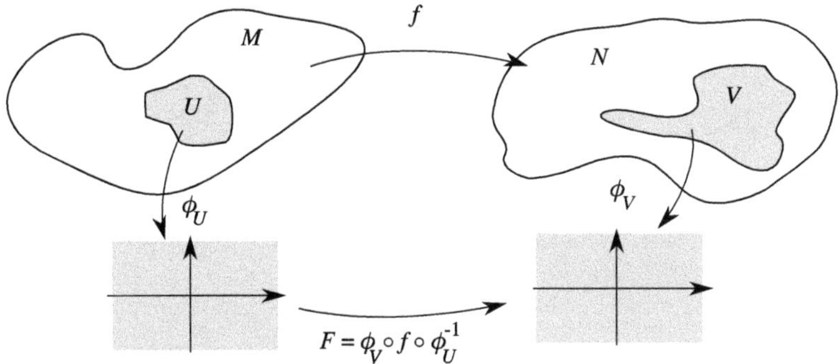

Fig. 2.17 A C^k function between two manifolds

$$\phi_{U_3}(x) = x, \ \ \psi_{U_3}(x) = 2x, \ \text{and } \theta_{U_3}(x) = \begin{cases} x & \text{if } x < 0 \\ 2x & \text{if } x \geq 0. \end{cases}$$

The atlases $\{\phi_{U_1}, \phi_{U_2}, \phi_{U_3}\}$ and $\{\psi_{U_1}, \phi_{U_2}, \psi_{U_3}\}$ are smoothly compatible, since $\psi_{U_3} \circ \phi_{U_3}^{-1} = 2x$ is smooth. However, the atlases $\{\phi_{U_1}, \phi_{U_2}, \phi_{U_3}\}$ and $\{\phi_{U_1}, \phi_{U_2}, \theta_{U_3}\}$ are not compatible, since the transition map $\theta_{U_3} \circ \phi_{U_3}^{-1}$ has a discontinuous derivative.

Definition 2.15 Each C^k atlas $\mathscr{U}$ on a topological manifold M is compatible with a unique largest C^k atlas, in the sense of inclusion. We call this atlas a *maximal atlas*. A topological manifold $(M, \mathscr{U})$ in which $\mathscr{U}$ is a maximal C^k atlas is called a C^k *manifold*. If $\mathscr{U}$ is a smooth atlas, then we call $(M, \mathscr{U})$ a *smooth manifold*, or simply a *manifold*. Often we will abbreviate notation by stating "M is a smooth manifold with atlas $\mathscr{U}$" instead of stating "$(M, \mathscr{U})$ is a smooth manifold."

Continuous functions between manifolds that respect the manifold structure are given special status.

Definition 2.16 Suppose that $(M, \mathscr{U})$ and $(N, \mathscr{V})$ are manifolds and that $f : M \to N$ is a continuous function. We call f a C^k *map* or C^k *function* if for each $U \in \mathscr{U}$ and $V \in \mathscr{V}$ for which $f(U)$ intersects V, $F = \phi_V \circ f \circ \phi_U^{-1}$ restricted to $(\phi_U \circ f^{-1})(V)$ has k continuous derivatives. This is well-defined because F is a function between open subsets of Euclidean space, as shown in Fig. 2.17. The set of C^k maps between M and N is written $C^k(M, N)$. If F has derivatives of all orders, we say f is a *smooth map* in $C^\infty(M, N)$.

C^k manifolds are important in applications because they support a notion of calculus. In particular, we can define a notion of a derivative for C^k maps.

Definition 2.17 Suppose $f : M \to N$ is a C^k function between C^k manifolds with maximal atlases $\mathscr{U}$ and $\mathscr{V}$, respectively. Given $x \in U \in \mathscr{U}$ and $V \in \mathscr{V}$, the

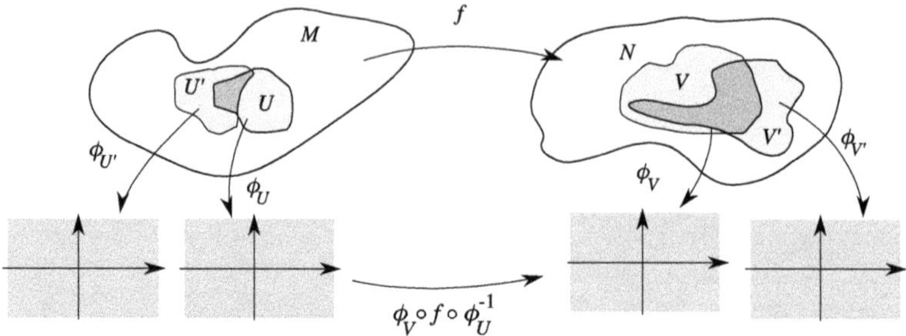

Fig. 2.18 Changes of charts on the domain and range of a C^k function between two manifolds

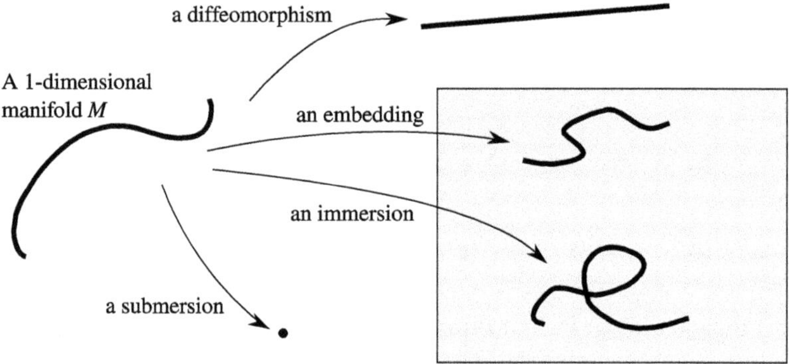

Fig. 2.19 Examples of different kinds of smooth maps of a 1-dimensional manifold

derivative of f at x is the Jacobian matrix of partial derivatives of $\phi_V \circ f \circ \phi_U^{-1}$. We will write $\mathrm{d}f_x$ for the derivative of f at x.

Lemma 2.1 *The rank of the derivative at a point $x \in M$ of a C^k function $f : M \to N$ between C^k manifolds is invariant under a change of charts.*

Proof Suppose that $\mathcal{U}$ and $\mathcal{V}$ are maximal C^k atlases of M and N, respectively. Observe that if x is also in $U' \in \mathcal{U}$ and $f(x)$ is also in V', then (see Fig. 2.18)

$$\begin{aligned}\phi_{V'} \circ f \circ \phi_{U'}^{-1} &= \phi_{V'} \circ \left(\phi_V^{-1} \circ \phi_V\right) \circ f \circ \left(\phi_U^{-1} \circ \phi_U\right) \circ \phi_{U'}^{-1} \\ &= \phi_{V'} \circ \phi_V^{-1} \circ \left(\phi_V \circ f \circ \phi_U^{-1}\right) \circ \phi_U \circ \phi_{U'}^{-1}\end{aligned}$$

The maps $\phi_U \circ \phi_{U'}^{-1}$ and $\phi_{V'} \circ \phi_V^{-1}$ are homeomorphisms between two Euclidean open sets and hence of full rank, a fact called the "invariance of dimension." Therefore $\phi_{V'} \circ f \circ \phi_{U'}^{-1}$ has the same rank as $\phi_V \circ f \circ \phi_U^{-1}$. □

Since dimension is an important local invariant of manifolds, maps that interact well with dimension have special names and properties as indicated in Fig. 2.19.

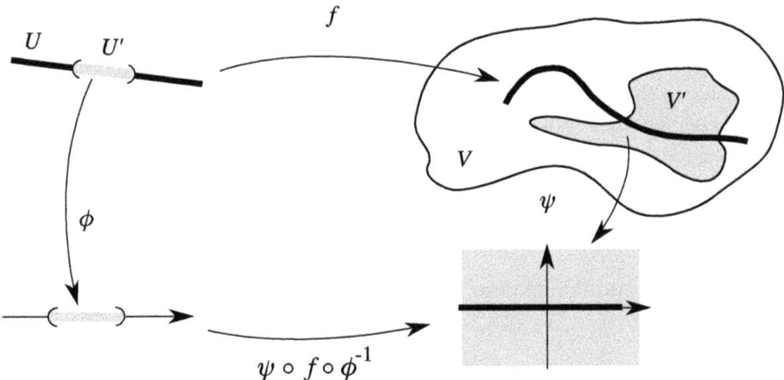

Fig. 2.20 A submanifold chart

Definition 2.18 Suppose $f : M \to N$ is a C^k function between C^k manifolds.

1. f is a *diffeomorphism* if it is a C^k homeomorphism with a C^k inverse. If f is a diffeomorphism, we say M and N are *diffeomorphic*.
2. If the derivative of f is injective at every point $x \in M$, then f is called an *immersion*.
3. An immersion that is also a homeomorphism onto its image is called an *embedding*.
4. If the derivative of f is surjective at every point $x \in M$, then f is called a *submersion*.

Written more simply, if a diffeomorphism exists between two manifolds, they are indistinguishable as manifolds. An embedding is a diffeomorphism into some ambient space. Immersions permit self-intersections. Submersions require the dimension of the domain to be greater than or equal to dimension of the range.

Definition 2.19 Suppose that N is a topological subspace of a manifold $(M, \mathscr{U})$. A *submanifold chart* for N is a chart $\phi : U \to \mathbb{R}^n$ of $\mathscr{U}$ in which $\phi(N \cap U)$ is the intersection of a linear subspace of $\mathbb{R}^n$ with $\phi(U)$. Explicitly, the *submanifold dimension* of N on U is the number k such that every point of the form $(x_1, \ldots, x_k, 0, \ldots, 0) \in \phi(U) \subseteq \mathbb{R}^n$ is the preimage of a point in N.

A submanifold $(N, \mathscr{V})$ of a manifold $(M, \mathscr{U})$ is a topological subspace of M which has an atlas $\mathscr{V}$ of submanifold charts.

Submanifold charts arise whenever there are C^k maps whose derivatives have constant rank.

Lemma 2.2 *(Loosely following (Lee 2003, Theorem 7.8)) Suppose $U \subseteq \mathbb{R}^m$ and $V \subseteq \mathbb{R}^n$ are open and $f : U \to V$ is a C^1 map whose derivative has rank k at every point $x \in U$. For any $x \in U$, there exist charts $\phi : U' \to \mathbb{R}^m$ and $\psi : V' \to \mathbb{R}^n$ with $U' \subseteq U$ and $f(U) \subseteq V' \subseteq V$ such that (see Fig. 2.20)*

$$\psi \circ f \circ \phi^{-1}(x_1, \ldots, x_m) = (x_1, \ldots, x_k, 0, \ldots, 0).$$

The lemma is essentially a nonlinear form of the singular value decomposition, since the derivative of f has rank k at x we have

$$\mathrm{d} f_x = A^{-1} \begin{pmatrix} I_{k \times k} & 0_{k \times m-k} \\ 0_{n-k \times k} & 0_{n-k \times m-k} \end{pmatrix} B.$$

But the Lemma is actually stronger; this equality can be made to hold *everywhere* in the chart U'.

Proof Without loss of generality, we can assume that $x = 0$, and that $f(0) = 0$. For convenience, we will construct ϕ and ψ so that $\phi(0) = 0$ and $\psi(0) = 0$. Without loss of generality assume that by permuting dimensions in both the domain and range, we can write $f(x_1, \ldots, x_m) = (g(x_1, \ldots, x_m), h(x_1, \ldots, x_m))$, where $g : U \to \mathbb{R}^k$ has full rank and $h : U \to \mathbb{R}^{n-k}$. Given that we want $\psi \circ f \circ \phi^{-1}$ to be the identity on the first k dimensions, let us define

$$\phi(x_1, \ldots, x_k, \ldots, x_m) = (g(x_1, \ldots, x_k, \ldots, x_m), x_{k+1}, \ldots, x_m).$$

We claim that ϕ^{-1} is defined, at least if ϕ is restricted to some smaller open set $U' \subset U$, by the classical inverse function theorem. This follows because

$$\mathrm{d}\phi_0 = \begin{pmatrix} A_{k \times k} & B_{k \times m-k} \\ 0_{m-k \times k} & I_{m-k \times m-k} \end{pmatrix}$$

is nonsingular since $(A_{k \times k} \; B_{k \times m-k})$ is the matrix of partial derivatives of g, which is of full rank by assumption. Therefore we have that on U',

$$f \circ \phi^{-1}(x_1, \ldots, x_m) = (x_1, \ldots, x_k, c(x_1, \ldots, x_m)),$$

where $c : U' \to \mathbb{R}^{n-k}$ is a C^k function. Thus, we would like to define ψ to be

$$(y_1, \ldots, y_k, y_{k+1} - c_{k+1}(x_1, \ldots, x_m), \ldots, y_n - c_n(x_1, \ldots, x_m)).$$

Perversely, though, this means that ψ isn't a map to the subset of $\mathbb{R}^n$ we wanted. However, we happen to be lucky since c is independent of $x_{k+1}, \ldots, x_m$. Observe that

$$\mathrm{d}(f \circ \phi^{-1}) = \begin{pmatrix} I_{k \times k} & 0_{k \times m-k} \\ A_{n-k \times k} & B_{k \times m-k} \end{pmatrix}$$

must have column rank exactly equal to k, since ϕ is a C^k diffeomorphism by construction. Therefore, $B_{k \times m-k}$ is actually a zero matrix, proving our claim that c is independent of $x_{k+1}, \ldots, x_m$.

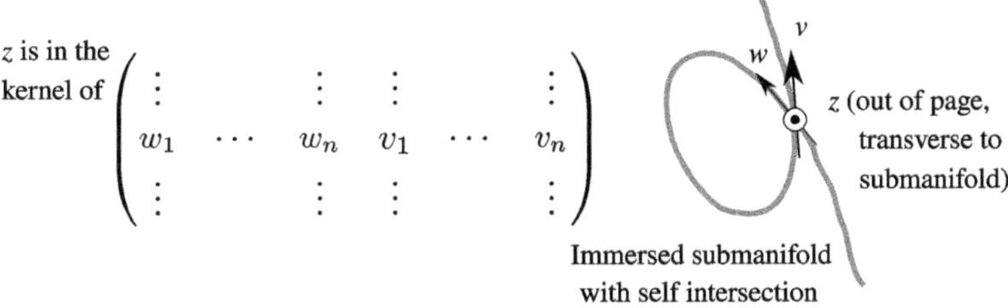

Fig. 2.21 Dimensional constraints in resolving self-intersections of an immersed manifold

Therefore, we have that $c(x_1, \ldots, x_k, x_{k+1}, \ldots, x_m) = C(x_1, \ldots, x_k)$, and so we may define

$$\psi(y_1, \ldots, y_n) = (y_1, \ldots, y_k, y_{k+1} - C_{k+1}(y_1, \ldots, y_k), \ldots, y_n - C_n(y_1, \ldots, y_k)).$$

□

This lemma facilitates the following proposition, that embeddings inject a copy of a manifold as a submanifold of another.

Proposition 2.2 *A C^k embedding $f : M \to N$ of one C^k manifold into another is also a C^k diffeomorphism onto its image $f(M)$, which is a submanifold of N.*

Proof Evidently, f is a homeomorphism when its range is restricted to $f(M)$. Since f is also an immersion, it is a map of constant (full) rank. Therefore, by the Lemma, we have that for every $x \in M$, there exist charts $\phi : U \subseteq M \to \mathbb{R}^m$ and $\psi : V \subseteq N \to \mathbb{R}^n$ so that $f(U) = f(M) \cap V$ (since f is a homeomorphism onto its image) and

$$\psi \circ f \circ \phi^{-1}(x_1, \ldots, x_m) = (x_1, \ldots, x_m, 0, \ldots, 0).$$

Hence ψ is a submanifold chart for $f(M)$ at $f(x)$. The inverse of f is C^k, since it is the identity with this choice of charts. □

The prototypical theorem about embeddings is the Whitney approximation theorem , which states that any compact manifold can be embedded in a Euclidean space of high enough dimension.

Theorem 2.1 *(Whitney approximation theorem (Whitney 1936, Theorem 2)) Suppose that $f : M \to \mathbb{R}^{2n+1}$ is a C^k map of a compact n-dimensional C^k manifold. Then for any $\epsilon > 0$, there exists a $g : M \to \mathbb{R}^{2n+1}$ such that $\|g(x)\| < \epsilon$ for every $x \in M$ and $f + g$ is an embedding. In this case, $\|v\|$ is the length of a vector v in $\mathbb{R}^{2n+1}$.*

The proof of Theorem 2.1 relies strongly on the (para) compactness of M, and a rather technical result called Sard's theorem. For an elementary proof, see Lee (2003).

More sophisticated versions involving transversality theory can be found in Golubitsky and Guillemin (1973). In contrast, an extremely elegant, elementary, and *constructive* approach is taken by Yomdin and Comte (2004) using something called "tame topology." In all cases, there are two essential aspects to the proof:

1. Showing that a C^k map can be perturbed[6] to a C^k immersion, and
2. Showing that a C^k immersion can be perturbed to a C^k embedding.

The dimensional limitation $\dim M \leq 2n + 1$ comes from the following observation. Assume that our map f has a self intersection, that $f(x) = f(y)$ for $x \neq y$. At worst case, the derivatives at x and y span a subspace of dimension $2 \dim M$ as shown in Fig. 2.21. The matrix of derivatives

$$\left(df_x \, df_y\right)$$

is a matrix of size $2n + 1 \times 2n$ and has rank at most $2n$. Thus, there exists at a vector $v \neq 0$ in the range of f that is not in the span of either derivative. We can move either x or y a small amount (bounded by ϵ) in the direction of v to remove the self intersection.

2.3 Case Study: Signal Manifolds for Localization, Tracking, and Navigation

When trying to "map out" an unknown environment, the locations of the measurements are of prime interest. *Remote imaging*, the process of inferring the location and properties of targets in a scene from measurements taken at a distance from the targets, has become an essential tool for cartographers. The measurements of a target need to be an unambiguous function of its position and other properties in order to be useful for locating and identifying it. Remote imaging requires *sources of illumination* or *transmitters* that provide signals that are scattered from targets in the scene and measured by *receivers*. We will assume that the transmitters and receivers are able to communicate amongst themselves without disrupting the collection process. This enables them to perform localization tasks using a smaller number of transmitters and receivers merely by moving them to different locations to form a "synthetic aperture."

Photographic techniques are often a preferred form of remote sensing due to their ease of interpretation. However, they suffer from many limitations due to weather, lighting conditions, and occlusions. For this reason, the use of lower frequencies both acoustically and electromagnetically has gained an indispensable role in cartography. Radio and sound waves can propagate long distances, which permits measurements to be taken far from the scene. In all of these cases, each location in the scene can be uniquely identified. For instance, positions in an optical image viewed by a

[6] A *perturbation* is a small change to a mathematical object.

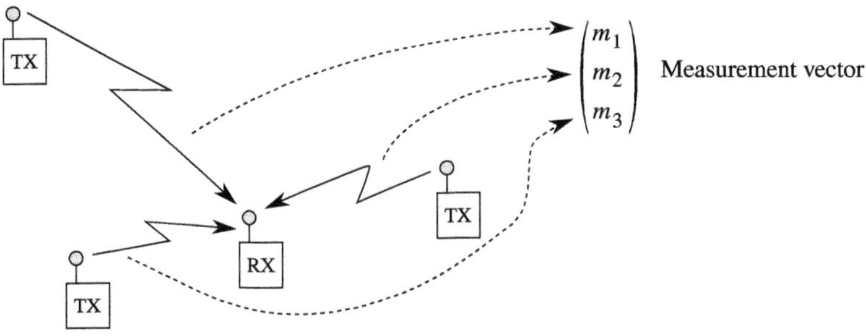

Fig. 2.22 Taking measurements of multiple signals at a receiver

camera or the human eye are discriminated by angle. (The accuracy of most optical sensors is usually quoted as an angular resolution.) Locations in a synthetic aperture radar image are discriminated by a unique distance to the sensor and associated Doppler frequency. If the signals in use do not travel in a straight line, it can seriously complicate interpretation, however. Magicians have often played with this ambiguity; the use of mirrors can result in an image for which the "easy" interpretation is incorrect and surprising. However, this ambiguity could be useful; reflective signals could permit imaging systems to "see around corners" (DARPA/STO 2010).

Remote imaging methods also play an important role in *navigation*, the method of inferring the location of a receiver in the scene. For instance, it is easy to see that measuring the distance from the receiver to three transmitters (in general position) whose locations are known suffices to determine the location of the receiver, if all are confined to move in a plane. What if the positions of the transmitters are not known, or the signal measurements do not correspond directly to distance? The Whitney approximation theorem provides a surprising answer: if the measurements of the signals depend smoothly on the receiver's location, then the receiver's location is completely determined using the signals from five (instead of three) transmitters. (Care must be exercised to ensure *transversality*—a generalization of general position—holds.)

Consider a setup as shown in Fig. 2.22. To each transmission link, the receiver takes a measurement, which is stored in a vector. To each receiver location, a different such vector is obtained, though we make the assumption that these vectors depend smoothly on the receiver location. Thus, all of the possible measurements can be encoded in a smooth map from the 2-dimensional plane (receiver locations) to a vector space whose dimension is the number of transmitters. The Whitney approximation theorem states that this map can be approximated by an injective map when the measurement dimension is more than twice the dimension of the plane, that is at least 5.

Realistic navigation systems based on this idea have additional complexity and redundancies. The most popular example is that of the Global Positioning System (GPS), in which transmitters are located on a constellation of 24 satellites (see Fig. 2.23) whose positions (as a function of time) are known very accurately. Each satellite transmits one of 12 coded signals; satellites with the same code have orbits

Fig. 2.23 A GPS satellite (image courtesy of US Air Force)

that place them on opposite sides of the earth. Therefore, every location on the earth is visible to at most 12 uniquely identifiable transmitters.

In addition to identity, the coded signals convey accurate timing information, which allows the receiver to synchronize its clock with the transmitter's clock. By measuring the apparent timing offsets between clocks associated to different transmitters, the receiver can measure its distance to each transmitter, thereby solving for its location. In order for this to succeed, the system of equations describing timing as a function of location must have a unique solution. This requires each location to have a unique collection of timing offsets.

Existing GPS systems also make the assumption that the signals propagate along straight lines, though they do account for propagation variations due to the Earth's gravitational field, relativity, and for average ionospheric conditions. This has an important implication: GPS does not work well in areas where reflective obstacles are present. For instance, consider the GPS track in Fig. 2.24 collected by the author near the University of Pennsylvania, in Philadelphia. The actual path was relatively straight (within a few feet), and remained on one side of the road. However, the GPS track shows erratic jumps in position, from one side of the road to the other. These jumps align fairly closely with the edges of nearby buildings. When he crossed 38th street (moving to the west), the buildings were much farther from the road and therefore diminished both the signal reflections and the GPS error. Because the track was not broken, we can infer that the author's position along this path was uniquely determined by the GPS signals. However, since there were occlusions and reflections, the GPS signals were no longer continuous (much less smooth) functions of his position. Therefore, the Whitney approximation theorem cannot be used in this

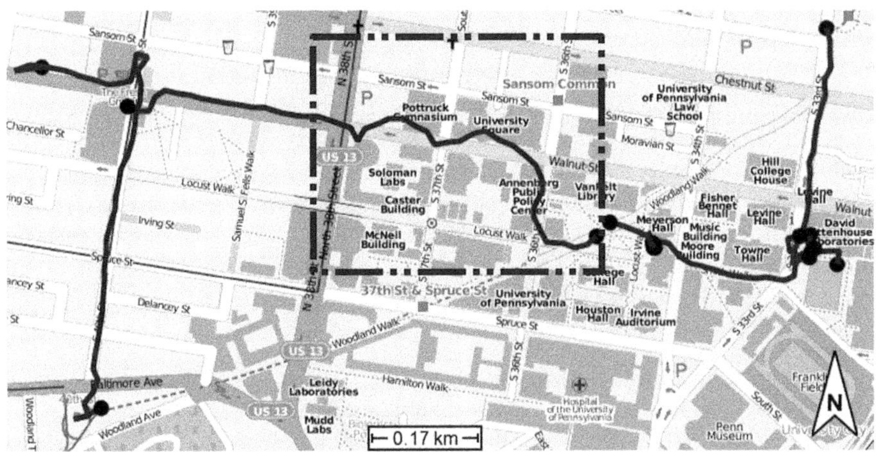

Fig. 2.24 An example GPS track collected by the author. East of 38th street, the buildings are close to the road and cause substantial error (*marked box*). West of 38th street, the buildings are far from the road so the error is reduced

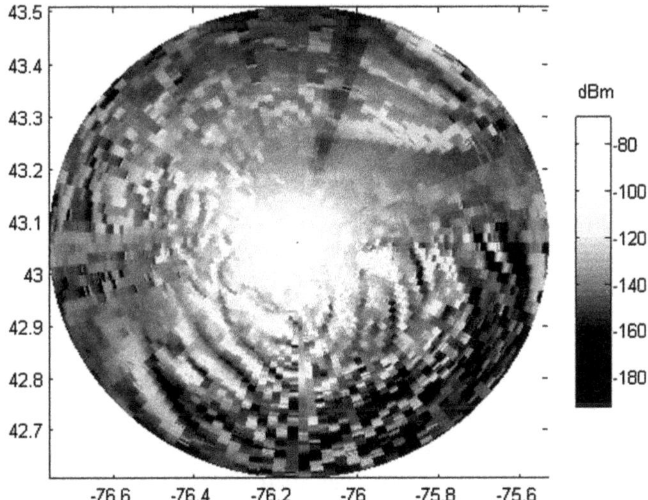

Fig. 2.25 The signal strength associated with a land-mobile radio communication tower located at the top of a hill varies dramatically with location. (Image produced using the model in Longley and Rice (1968) as implemented by the author, in collaboration with SRC, Inc.)

situation. However, a generalization of the Whitney approximation theorem called the Signal embedding theorem (Theorem 2.2), indicates that the receiver's location is still uniquely determined.

2.3.1 Signal Manifold Fingerprinting

Suppose that the propagation environment is a manifold M, and that there are n transmitters $T = \{t_1, \ldots, t_n\} \subset M$, and m receivers $R = \{r_1, \ldots, r_m\} \subset M$. Each transmitter emits a signal whose properties can vary over some portion of M. For instance, its signal strength, first time of signal arrival, polarization, antenna orientation, or bit error rate can vary dramatically according to various physical conditions as shown in Fig. 2.25. In order to provide substantial theoretical and practical flexibility, we take a very general approach to the received signal. Therefore, for each transmitter t_i, we assign

1. An open submanifold with compact closure $U_i \subseteq M$ describing the *coverage region* over which the signal can be reliably received,
2. A *signal manifold* S_i describing the possible parameters of the signal, for instance its signal strength, direction of arrival, or polarization, and
3. A smooth *propagation function* $P_i : U_i \to S_i$, which describes the received signal at every location in U_i.

Taken together, all of the transmitters generate a *transmission profile* that indicates all of the received signals at every given location. Formally, we extend each signal manifold with an additional point $\perp$, which indicates failure to receive that transmitter. Thus each propagation function extends to $P_i : M \to S_i \sqcup \perp$, which is likely to be discontinuous. We then collect all of these signal manifolds into a product, making the transmission profile a function $P : M \to \prod_{i=1}^{n}(S_i \sqcup \perp)$.

Unfortunately, the Whitney approximation theorem cannot be used on transmission profiles for two main reasons:

1. The overall collection of signals is discontinuous as a receiver crosses the boundary of a coverage region.
2. The signal perturbations associated to different transmitters are independent of one another.

However, a variant of the Whitney approximation theorem, called the signal embedding theorem (Theorem 2.2) does apply. In this more general setting, the result is somewhat weaker and does not imply that the transmission profile is an embedding in the sense of manifolds. However, it allows one to assert that an arbitrarily small perturbation of a transmission profile can be found which is injective and locally an embedding. The discontinuities of the perturbed transmission profile then result in a disconnected image in the signal manifold of a connected domain. In order to prove this result, we need a few preliminaries from transversality theory (Lee 2003, Golubitsky and Guillemin 1973).

Since the number of transmitters that each receiver can detect varies as a function of position, we require a bound on this number.

Definition 2.20 For a transmission profile P, the smallest number of non-$\perp$ components of $P(x)$ over all of M is called the *depth* of P, $\operatorname{dep} P$. (See the left frame of Fig. 2.26.)

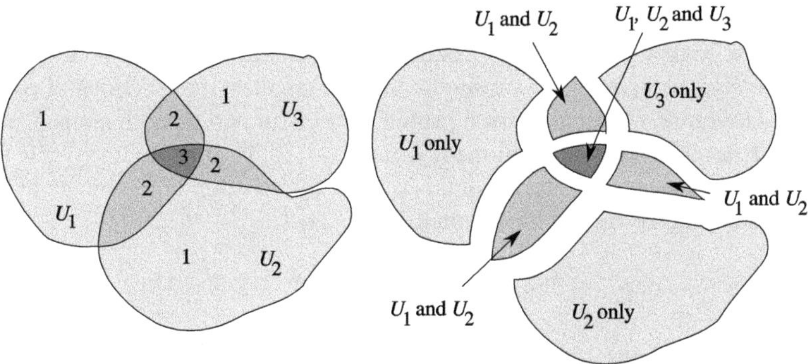

Fig. 2.26 Three coverage regions U_1, U_2, U_3 with mutual intersections. The resulting depth as a function of receiver location (*left*) and receiver locations partitioned according to transmitter coverage (*right*). The total depth in this example is 1

Definition 2.21 Suppose M and N are C^k manifolds and that $f : M \to \mathbb{R}^n$, $g : N \to \mathbb{R}^n$ are immersions. We say that f and g are *transverse at a point* x if their images intersect at x and the matrix of derivatives

$$\left(\mathrm{d}f_x \; \mathrm{d}g_x\right) \tag{2.2}$$

is of full rank.

Proposition 2.3 *If $f : M \to \mathbb{R}^n$ and $g : M \to \mathbb{R}^n$ are transverse at x, their images intersect in a submanifold of dimension $n - \dim M - \dim N$.*

Proof This result is an immediate consequence of the dimension theorem in linear algebra using two submanifold charts as constructed by Lemma 2.2. □

When the dimension $n - \dim M - \dim N < 0$, the method used to prove Theorem 2.1 (Whitney approximation) shows that a small perturbation of either map will result in no intersection at all. Stated more precisely, the set of maps in the space $C^k(M, \mathbb{R}^n)$ that do not intersect the image of g is open and dense (Golubitsky and Guillemin 1973). More generally, the set of maps in $C^k(M, \mathbb{R}^n)$ that are transverse to g is always open and dense when M is compact.

Theorem 2.2 *(Signal embedding theorem (Robinson and Ghrist 2012, Theorem 3)) There is an open and dense set of propagation functions for which the associated transmission profile is injective if $2 \dim M < \operatorname{dep} P$.*

Proof (sketch) Although the proof is fairly technical, the idea is simple: apply the Whitney approximation theorem on subsets of M illuminated by the same set of transmitters (right frame of Fig. 2.26). Points in different such subsets necessarily have different transmission profiles, and so cannot spoil the injectivity of the profile.

The technical complexity of the proof comes from showing that the transmission profile is injective on an open and dense set of propagation functions *independently* of one another; we sketch the argument here. Suppose that the transmission profile is not injective, so that $P(x) = P(y)$. Take open neighborhoods V and W around x and y respectively, each contained entirely within the smallest intersections of coverage regions. Therefore, the transmission profile restricts to two smooth maps $f = P|_V$ and $g = P|_W$. The set of smooth maps from $V \to \mathbb{R}^{\text{dep}P}$ that are transverse to g is open and dense, as noted before. On this set, the intersection of the image of such a map and the image of g will have dimension at most

$$\text{dep}P - \dim V - \dim W = \text{dep}P - 2\dim M,$$

since the image of the transmission profile has dimension at most dep$P$. If this intersection dimension is less than zero, the set of maps whose image do not intersect the image of g will be open and dense. □

2.3.2 Multiple Target Detection and Localization

All forms of remote imaging rely on a unique signal response from each target in the scene; as a result, several kinds of systems are applications of the signal embedding theorem. As discussed previously, GPS is a popular navigation system that relies on straight-line distances to satellite transmitters. Consider the satellite transmitters with locations $t_1, \ldots, t_{12}$, and the receiver with location r. If there are no reflections and the transmitters are synchronized to one another, the time of arrival of each signal at r is $\tau_i = d(t_i, r)/c$ where c is the speed of light and $d(\cdot, \cdot)$ computes the distance between two points (corrected for relativistic effects). Since the GPS receiver does not have an accurate absolute time reference, it computes differences between the first time of arrival and all the others. Therefore, the transmission profile is a smooth map that takes $r \mapsto (\tau_2 - \tau_1, \ldots, \tau_{12} - \tau_1)$.

The standard algorithm assumes that the signals propagate along geodesics, and this map can be inverted explicitly. However, if six satellites are always in view, then the transmission profile has depth 5. According to the signal embedding theorem, the transmission profile will be injective on the 2-dimensional surface of the earth, *even if some of the signals have been reflected or occluded*. If instead, seven satellites are always in view, then the transmission profile can be used to discriminate between receivers with the same latitude and longitude but different heights.

One could also take the transmission profile to be $(r, t_1) \mapsto (\tau_2-\tau_1, \ldots, \tau_{12}-\tau_1)$, which encodes the desire to solve for synchronization as well as location. In this case, eight satellites could be used to solve for 3-dimensional position and absolute time, again even if the signals have been reflected off buildings. At present, no GPS receivers use this idea, as it would apparently require tabulating or computing the expected reflections from known buildings. However, the signal embedding theorem is important because it places *lower* bounds on what is possible.

Taking the GPS idea further, one can imagine a radar system called a *moving target indicator*, which tracks a number of targets (Stimson 1998). In this case,

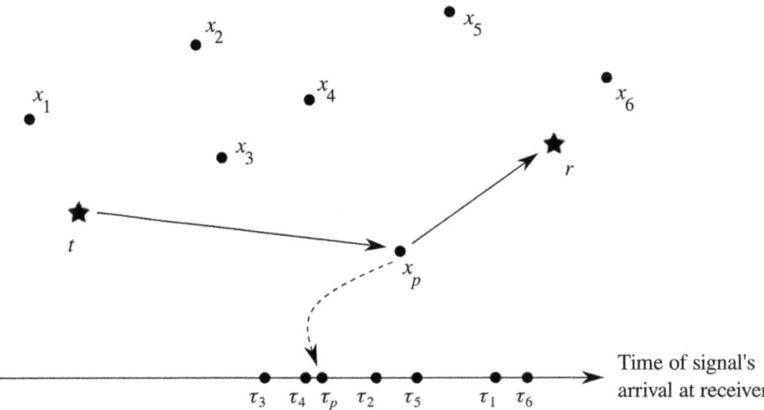

Fig. 2.27 Multiple targets being tracked by a transmitter and receiver

suppose there is a single transmitter t, receiver r, and a number of targets $x_1, \ldots, x_p$ as shown in Fig. 2.27. We can safely assume that the transmitter and receiver have known locations and are synchronized, but that the scene may have complicated reflective and obstructive geometry. Assume that the first time a signal traverses the path $t \to x_i \to r$ is given by the propagation function $\tau_i(x)$, which is smooth on some open set of positions for the target $U_i \subseteq M$. (Different targets may be more or less visible, so τ_i and τ_j may be different.) Outside of U_i, τ_i takes the value $\perp$.

We can then assemble a transmission profile from the τ_i, which has a depth of at most p. However, since there are p targets, the domain of the transmission profile has dimension $p \dim M$, which is too large for the signal embedding theorem. In this case, the signal embedding theorem is indicating that the problem is underdetermined. We can supply additional information in a variety of ways. For one, we can augment τ_i with an additional dimension by measuring the signal strength of the response. However, experience with radar and sonar systems suggests that this is somewhat unreliable, due to substantial variations in scattering cross section (Stimson 1998).

Instead, the usual way to solve the problem is to add multiple transmitters or receivers (or both) to supply additional measurements. If we add additional transmitter-receiver pairs for a total of q pairs and construct them so that they do not interfere, then the depth could be as large as pq. Thus, in the best case scenario $q > 2 \dim M$ would suffice to uniquely solve for each target's location, again without constraints on the geometry.

Having multiple transmitters and receivers is often expensive, so many remote sensing systems utilize moving sensors to form a synthetic apertures, see Jakowatz et al. (1996), Carrara et al. (1995). (See Fig. 2.28 for an example image product.) They usually rely on the choice of a coded waveform that permits the correlation of temporal characteristics against other properties of the scene. In this way, if a unique location can be extracted from the timing alone, one can then use the location to extract other properties. Specifically, consider the case of a moving transmitter $t(\tau)$

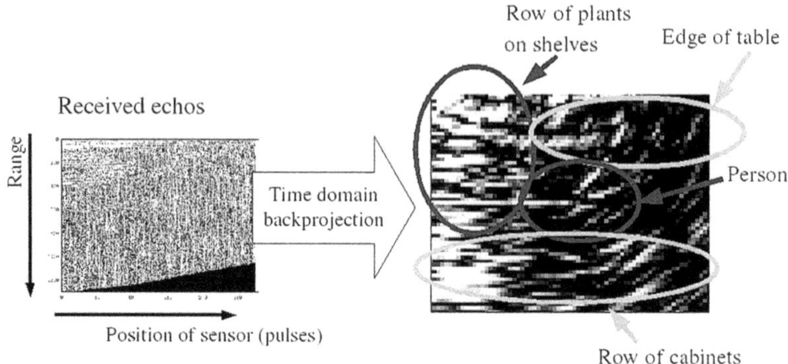

Fig. 2.28 An example synthetic aperture sonar image of the author's kitchen

and receiver $r(\tau)$. If M has a metric d on it, then we can write the range (distance from transmitter, to target, to receiver) and Doppler frequency (rate of change of range) explicitly. For location $x \in M$ in the scene, one obtains a range (from arrival times)

$$R(\tau, x) = d(t(\tau), x) + d(x, r(\tau)),$$

and a range rate (from the Doppler frequency offset)

$$D(\tau, x) = \frac{\partial}{\partial \tau} R(\tau, x).$$

Observe that the map $x \mapsto (R(\tau_1, x), D(\tau_1, x), \dots, R(\tau_p, x), D(\tau_p, x))$ is a signal profile whose depth is $2p$. Hence, if $p > \dim M$, the signal embedding theorem ensures that we can discriminate locations (and hence material properties at those locations) based on timing alone.

2.4 Open Questions

1. The signal embedding theorem, while helpful from a feasibility point of view, is unfortunately nonconstructive. If it states that the transmission profile is injective, it provides no insight about how to implement a system to exploit the profile. The obvious implementations of the applications we have outlined require substantial tabulation of previous measurements, which is probably not ideal. At present, no constructive algorithms exploit the transmission profile directly.
2. Within each coverage region, the propagation function is smooth and perhaps therefore somewhat uninformative. Traditional methods focus on propagation functions, because they can be completely characterized. This is a weakness if there are many reflections and occlusions. On the other hand, the boundaries of

the coverage regions are both well-defined and lower dimensional. Crossing a shadow boundary therefore conveys substantial location information. Can one organize the shadow boundary information more efficiently or effectively than the individual propagation functions?

3. If there is a sufficiently dense set of reflective targets, these coverage region boundaries will be rather prevalent. However, the coverage regions are now dependent on locations of scatterers as well as the locations of the transmitters and receivers, and several coverage regions may arise from each illumination. What geometric bounds arise are there on the prevalence of shadows?
4. The collection of intersections of coverage regions forms a cell complex, in which the cells are labeled with which transmitters are detectable. Are there constraints on how typical faces of these cells could be labeled? Are there statistical or asymptotic properties that should be expected when the number of cells is large?
5. Traditional image formation methods require careful synchronization between transmitter and receiver, as assumed in our discussion above. If relaxed, one can construct a transmission profile to solve for the unknown time offsets. However, it is unclear what algorithms are possible in this context. Specifically, is there a general algorithmic formulation of synthetic aperture image formation that is not dependent on knowledge of the collection geometry and timing?

References

Banyaga A, Hurtubise D (2004) Morse homology. Springer, Berlin

Carrara W, Goodman R, Majewski R (1995) Spotlight synthetic aperture radar: signal processing algorithms. Artech House, Norwood

DARPA/STO: MER data collection review (2010). http://www.darpa.mil/STO/solicitations/baa09-01/presentations/MER_Data_Review.pdf

Golubitsky M, Guillemin V (1973) Stable mappings and their singularities. Springer, New York

Jakowatz C, Wahl D, Eichel P, Ghiglia D, Thompson P (1996) Spotlight-mode synthetic aperture radar: a signal processing approach. Kluwer Academic Publishers, Dordrecht

Kaczynski T, Mischaikow K, Mrozek M (2004) Computational homology. Springer, New York

Lee J (2003) Smooth manifolds. Springer, New York

Longley A, Rice P (1968) Prediction of tropospheric radio transmission loss over irregular terrain. A computer method. Technical Report. Boulder, CO

Milnor J (1963) Morse theory. Princeton University Press, Princeton

Robinson M, Ghrist R (2012) Topological localization via signals of opportunity. IEEE Trans Sig Proc 60(5):2362–2373

Steen L, Seebach JA (1978) Counterexamples in topology. Dover, New York

Stimson GW (1998) Airborne radar. SciTech, Raleigh

Whitney H (1936) Differentiable manifolds. Annal Math Second Ser 36(3):645–680

Yomdin Y, Comte G (2004) Tame geometry with application in smooth analysis. Springer, Heidelberg

Chapter 3
Signals

This chapter will explain that

1. Signals can be encoded in *cellular sheaves*,
2. *Sections* of cellular sheaves capture the notion of local consistency,
3. Signal processing systems can be represented as *sequences* of cellular sheaves called *topological filters*, and
4. Processing algorithms which exploit a sheaf model of signals can be more bandwidth-efficient than traditional approaches.

Topological spaces are the stage on which signal processing plays. Based on the tools developed in Chap. 2, it is advantageous to parametrize a collection of sensors by a topological space. Sensors, however, do not exist merely to be related to one another. Rather, they collect additional information about their environment. This information should be aggregated so that inferences about the environment can be made. This chapter outlines what form these *measurements* can take, how they are organized, and several ways they can be manipulated.

The perspective of most sensors is local; they only measure their immediate environment, or a very small portion of the wider environment. However, there is usually an expectation of consistency between nearby sensors. This consistency facilitates the comparison of measurements, as well as the process of drawing inferences from them. The mathematical structure used to organize local information and expose it conveniently for inferences is the *sheaf*.

3.1 Locality: Principles and Axioms

The central object in the study of topological signal processing is the *sheaf*, which specifies the totality of possible measurements collected by a family of sensors. We will usually use the conceptual model of a *vector space* to represent the range of measurements collected by an individual sensor, and therefore will focus most of our attention on sheaves of vector spaces. In almost all cases of interest, this level

M. Robinson, *Topological Signal Processing*,
Mathematical Engineering, DOI: 10.1007/978-3-642-36104-3_3,

of generality is both sufficient to describe the phenomenon being measured, but not too general to be awkward. Occasionally, we will have a need for a more general class of measurements, such as sets or groups.

If information is local, it is only valid within some region. For instance, it may be local in space, time, or frequency. Locality is described by two central principles:

1. Restricting attention from a larger region to a smaller region corresponds to a well-defined way to restrict the information collected from the larger region to the smaller region, and
2. If the information over two overlapping regions is mutually consistent, then the information should be valid over the union of those two regions.

Therefore, our mathematical definition of a sheaf will follow both of these principles. However, in order to facilitate a more computationally-feasible formulation, we represent *larger* regions by *lower* dimensional cells of a cell complex. This representation foreshadows the definition of a *nerve* of an open cover (Definition 4.17).

Definition 3.1 (following Shepard (1980)) A *sheaf of vector spaces* $\mathscr{S}$ (or a *sheaf* or a *signal*) on a cell complex X (called the *base space* of $\mathscr{S}$) is the assignment of

1. A vector space $\mathscr{S}(c)$ to each cell c of X. This vector space is called the *stalk* at c.
2. A linear map $\mathscr{S}(a \rightsquigarrow b) : \mathscr{S}(a) \to \mathscr{S}(b)$ to each attachment $a \rightsquigarrow b$ of a higher dimensional cell b to a lower dimensional one a, called the *restriction of a to b*

such that

1. the restriction of a to itself is the identity map, and
2. if $a \rightsquigarrow b$ and $b \rightsquigarrow c$, then $\mathscr{S}(b \rightsquigarrow c) \circ \mathscr{S}(a \rightsquigarrow b) = \mathscr{S}(a \rightsquigarrow c)$.

In a similar way, a sheaf of groups assigns a group to each cell, and a group homomorphism to each attachment. Sheaves of sets can be defined similarly, without requiring the restrictions to be anything other than functions.

More plainly, a sheaf is an object that "lives" over a cell complex. The part of the sheaf associated to a particular cell is called a *stalk* over that cell. If two cells are attached, their stalks are related by a *restriction*, which is a function from the stalk over the lower dimensional cell to the stalk over the higher dimensional cell.

Sheaves are quite flexible mathematical data structures. In order to highlight the diversity of sheaves that is possible, and the diversity of their applications, several significantly different examples are discussed below. Because of this diversity, and because of the abstractness of the definition, it is remarkably helpful to visualize sheaves in terms of the attachment diagram associated to the base space. *The diagram of restriction maps between stalks is the same as the attachment diagram.*

Example 3.1 Each school that a student attends maintains records of that student's grades, which are only shared in a limited fashion. As a simplified example of this process, consider a student who has attended high school, an undergraduate institution, a graduate institution, and then takes a postdoctoral position. We represent each institution as a vertex in a cell complex, as shown in Fig. 3.1. Every pair of institutions

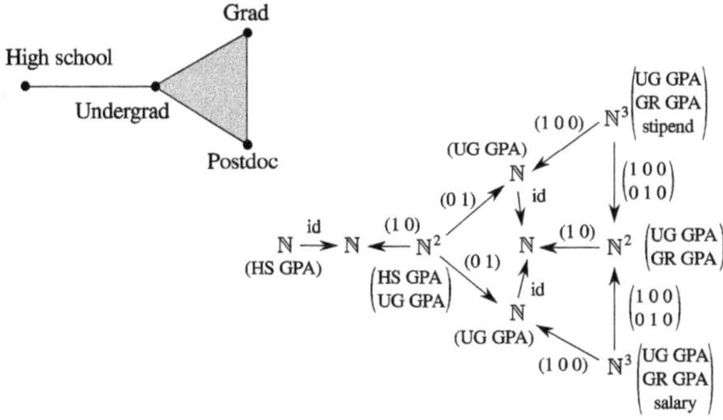

Fig. 3.1 A network of academic institutions that might share information about a student (*left*), and a sheaf representing pertinent information about a single student (*right*)

that share a piece of information is represented as an edge between their respective vertices. A common piece of information that is shared among three institutions is represented as a 2-simplex. For instance, high schools typically only communicate with undergraduate institutions, so there are no edges between a high school's vertex and any other institutions.

Let us assume the following:

1. The high school only keeps a record of the high school GPA,
2. The undergraduate institution keeps records of both the high school GPA, and the undergraduate GPA,
3. The graduate institution keeps records of the undergraduate GPA, graduate GPA, and any graduate stipend,
4. The postdoctoral institution keeps records of the undergraduate GPA, graduate GPA, and postdoctoral salary,
5. Stipend and salary information is not shared between institutions,
6. Grades are shared as appropriate, and should be consistent.

These assumptions lead to the sheaf structure shown on the right of Fig. 3.1. Each piece of information is represented by a natural number (grades and salaries cannot be negative, and are rounded to the nearest whole number). In this sheaf structure, the stalk over each vertex contains the information held by each institution. Each edge of the complex contains the information shared by the two institutions. Each 2-simplex contains the common information among three institutions, which in this example is only the undergraduate GPA. Each restriction map is represented by a projection matrix that selects the appropriate information to be shared. In particular, the restriction maps from the two postgraduate institutions do not share any information regarding the student's pay.

Unlike the previous example, the stalks and restriction maps might all be the same. As we will see in Chap. 6, these *constant sheaves* are useful for studying the topological structure of cell complexes, rather than any information stored "on" them.

Example 3.2 Consider the following abstract simplicial complex structure X for $\mathbb{R}$: $X^0 = \mathbb{Z}$ and $X^1 = \{(n, n+1)\}$. This is called the *usual simplicial structure for* $\mathbb{R}$. The easiest sheaf to construct from a given vector space V is the *constant sheaf* $\mathscr{V}$, which assigns the stalk $\mathscr{V}(c) = V$ to each cell c, and assigns identity maps to each attachment as the restrictions. For instance, the attachment diagram associated to X is

$$\cdots \leftsquigarrow \{n-1\} \rightsquigarrow \{(n-1, n)\} \leftsquigarrow \{n\} \rightsquigarrow \{(n, n+1)\} \leftsquigarrow \cdots \tag{3.1}$$

The constant sheaf on (3.1) is therefore given by the following diagram

$$\cdots \xleftarrow{\mathrm{id}_V} V \xrightarrow{\mathrm{id}_V} V \xleftarrow{\mathrm{id}_V} V \xrightarrow{\mathrm{id}_V} V \xleftarrow{\mathrm{id}_V} \cdots$$

The concept of a sheaf is also flexible enough to allow neighboring cells to have no consistency requirement.

Example 3.3 A different sheaf on the abstract simplicial complex (3.1) for $\mathbb{R}$ is the *V-sampling sheaf supported on* $\mathbb{Z} \subset \mathbb{R}$. It has the diagram

$$\cdots \longleftarrow V \longrightarrow 0 \longleftarrow V \longrightarrow 0 \longleftarrow \cdots$$

This sheaf structure will be useful in describing sampling theory (Sect. 4.5) in which measurements are taken at discrete locations or times with no measurements in between. In this particular situation, no consistency is maintained between neighboring measurements because the restriction maps are all zero.

More generally, a *V-sampling sheaf* supported on a subset A of a cell complex X is a sheaf whose stalks are V on each cell in A and 0 elsewhere. The restrictions between two cells $a \rightsquigarrow b$ both in A are identity maps, but all other restrictions are zero maps.

Sheaves are well-suited to represent timeseries, images, and video. In each of these data types, it is reasonable to suspect that the information they contain is concentrated locally. The V-sampling sheaf is a convenient way to represent discrete timeseries signals. The next example shows how to represent continuous timeseries using a related construction.

Example 3.4 Suppose that $C((a, b))$ is the vector space of continuous functions on the open interval (a, b). Then the following diagram describes the *sheaf of continuous functions* over the real line

$$\cdots \longrightarrow C((-1, 0)) \longleftarrow C((-1, 1)) \longrightarrow C((0, 1)) \longleftarrow \cdots$$

in which the arrows represent restricting the domain of a continuous function. For instance, the arrow $C((-1, 1)) \to C((0, 1))$ takes the function $f = f(x)$ defined on $(-1, 1)$ to the function which takes the same values on $(0, 1)$, but is not defined

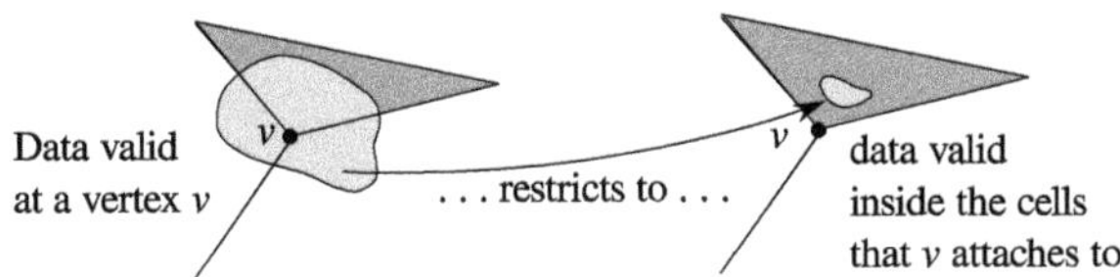

Fig. 3.2 The stalk of a sheaf should be thought of as specifying the data valid in an open set containing a cell

outside that interval. In this way, the data stored over each cell is merely a part of a function defined on a potentially much larger domain.

Exercise 3.1 Extend Example 3.4 to represent grayscale images, which are continuous functions over the plane $\mathbb{R}^2$. Use the 2-dimensional cell complex structure in which vertices are points (x, y) with integer coordinates, edges are length 1 vertical and horizontal line segments, and 2-cells are open squares with side length 1.

Remark 3.1 One should think of the stalk of a sheaf at a cell c as specifying the kind of information that could be present in an open neighborhood containing c. Open sets at a vertex therefore have a large impact in how information encoded in a sheaf is related to information over the higher dimensional cells connected to that vertex, as shown in Fig. 3.2. In this way, sheaves satisfy principle (1) of local information. In order to see how sheaves satisfy principle (2), we will need to wait until the definition of a *section* in Definition 3.3. This principle leads to the treatment of sheaves over posets in Baclawski (1975), which is similar in structure to the cellular theory discussed in this book.

The reader may find a comparison with more traditional treatments enlightening. Both Iverson (1984) and Bredon (1997) are good introductions to the general theory of sheaves, although the author sometimes finds the older Godement (1958) to be helpful. Because these three texts treat sheaves over general topological spaces (as opposed to sheaves over cell complexes), they must manage additional technical complexity. A more intuitive treatment can be found in the appendix of Hubbard (2006), which uses the Čech construction. The Čech construction defines sheaves over a lattice of open covers, and has a combinatorial feel.

The next example shows how sheaves can be used to represent piecewise linear functions on a graph. The sheaf structure allows the functions to have abrupt changes in slope which occur at each vertex.

Example 3.5 Suppose G is a graph in which each vertex has finite degree (G can be realized as an abstract simplicial complex). Assign arbitrary directions to each edge in G, so that it has the structure of a directed graph. (The directions are used for accounting purposes only, see Exercise 3.4). Let $\mathscr{PL}$ be the sheaf constructed on G that assigns $\mathscr{PL}(v) = \mathbb{R}^{1+\deg v}$ to each vertex v and $\mathscr{PL}(e) = \mathbb{R}^2$ to each edge e. (See Fig. 3.3). To each attachment of a degree k vertex v to an edge e, let $\mathscr{PL}$ assign the linear function $\mathscr{PL}(v \rightsquigarrow e) \colon \mathbb{R}^{1+k} \to \mathbb{R}$ given by

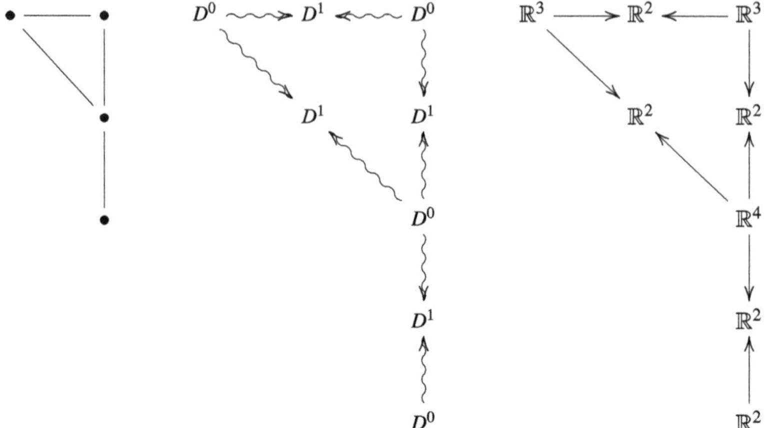

Fig. 3.3 Example of a graph (*left*), its attachment diagram (*center*), and sheaf $\mathscr{PL}$ (*right*)

$$\mathscr{PL}(v \rightsquigarrow e)(y, m_1, ..., m_e, ..., m_k) = \left(\frac{1}{2}m_e \pm y, m_e\right)$$

where we take the plus sign if e is pointing *inward* to v or the minus sign if e is pointing *outward* from v. $\mathscr{PL}$ is the sheaf of *piecewise linear functions* on G.

Exercise 3.2 Explain why $\mathscr{PL}$ (from Example 3.5) is the sheaf of piecewise linear functions by constructing it on the usual simplicial structure for $\mathbb{R}$. Hint: what information is necessary to define a line segment from $x = n$ to $x = n+1$ for a given integer n?

Sheaves can model physical situations, including flows of commodities over networks. These *flow sheaves* are based on the idea of conservation laws and provide numerous valuable examples.

Example 3.6 Suppose G is a directed graph in which each vertex has finite indegree and outdegree. Let one vertex be labeled $*$, which represents the connection of the network to the outside world. We call this $*$ the *external vertex*. We can represent the flow of a commodity (Ghrist 2011) over this graph as a *flow sheaf* of $\mathscr{F}$ that takes values in $\mathbb{N}$. Each value specifies an amount of commodity in transit at a particular location on the graph. We construct the sheaf according to the following specification (see Fig. 3.4)

1. The stalk over each edge is $\mathbb{N}$,
2. The stalk at the external vertex satisfies $\mathscr{F}(*) = \mathbb{N}^{\text{indeg}\,*+\text{outdeg}\,*}$ (commodities are *not* conserved at $*$),
3. For any other vertex, the stalk $\mathscr{F}(v) = \{(a_1, \ldots, a_m, b_1, \ldots, b_n) \in \mathbb{N}^{m+n} : \sum_{i=1}^{m} a_i = \sum_{j=1}^{n} b_j\}$, where m is the indegree of v and n is the outdegree, and
4. The restrictions $\mathscr{F}(v \rightsquigarrow e)$ are projections onto the component corresponding to the edge e.

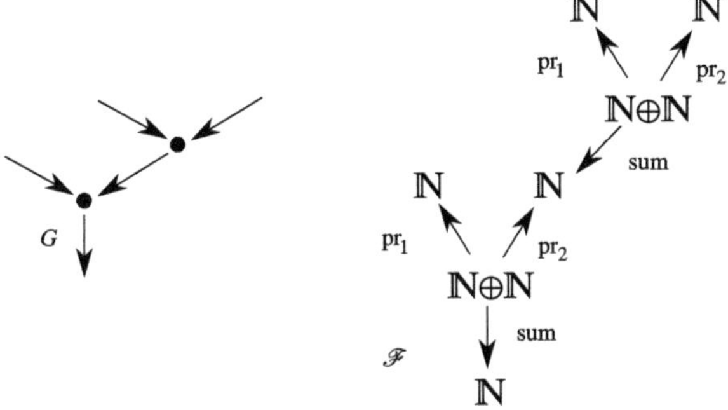

Fig. 3.4 An example of a flow sheaf $\mathscr{F}$ (*right*) on a directed graph G (*left*)

(It is often actually easier to manipulate values in a ring such as $\mathbb{Z}$ or field such as $\mathbb{R}$, but these include *negative* or *fractional* amounts of commodities, which may not be physically appropriate.)

Exercise 3.3 Construct flow sheaves on each of the following directed graphs. Be sure to specify both the stalks and the restrictions.

1.

2.

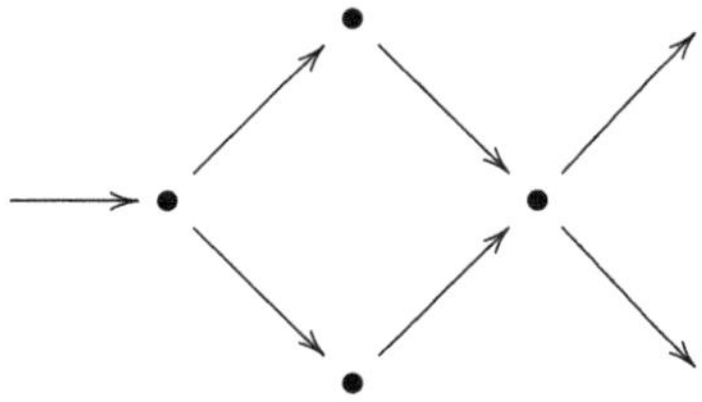

3.

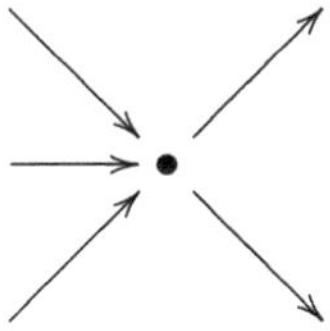

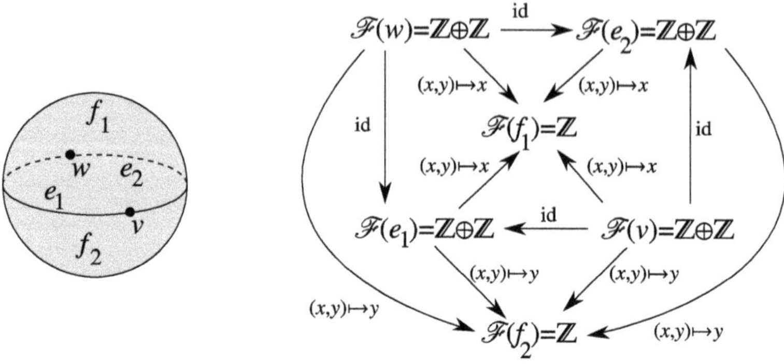

Fig. 3.5 A cell complex structure on the 2-dimensional sphere (*left*) and a sheaf over it (*right*)

4.

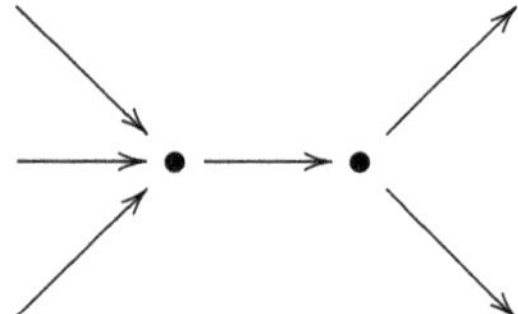

Example 3.7 Sheaves can be defined over higher dimensional cell complexes. Consider the sheaf structure shown in Fig. 3.5. An important property of the diagram of the sheaf is that it is *commutative*, since there are three ways that attach a vertex v to the 2-cell f_1, (directly, via e_1, and via e_2) each of these ways must yield the same function $\mathscr{F}(v) \to \mathscr{F}(f_1)$.

Remark 3.2 The algebraic and topological structure of the stalks is usually best suggested by the problem under study. However, it is worth noting that sometimes more algebraic structure than apparently necessary can yield important information. The process of enriching the algebraic structure of a sheaf is called *categorification* and is a fruitful way to obtain better indications of the behavior of a system. While we will not attempt a systematic treatment of categorification, the interested reader can find an example in Robinson (2012).

3.1.1 Sheaf Morphisms

A *sheaf morphism* is a consistent way to translate data in one sheaf into another. We begin by defining a morphism between sheaves on a fixed space X, and then later (Definition 3.4) extend the definition to handle sheaves on different spaces as well.

Definition 3.2 A *morphism* $f : \mathscr{S} \to \mathscr{R}$ of sheaves on a cell complex X assigns a linear map $f_a : \mathscr{S}(a) \to \mathscr{R}(a)$ to each cell a so that for each attachment $a \rightsquigarrow b$, the compatibility condition $f_b \circ \mathscr{S}(a \rightsquigarrow b) = \mathscr{R}(a \rightsquigarrow b) \circ f_a$ holds. Sheaf morphisms therefore commute with the restrictions of $\mathscr{R}$ and $\mathscr{S}$. If the sheaves are not of topological vector spaces, we require that the functions f_a be the appropriate structure-preserving map for the kind of sheaves in question. For instance, homomorphisms are used for sheaves of groups. A *sheaf isomorphism* is a sheaf morphism in which each of the f_a are isomorphisms.

It is sometimes helpful to visualize the definition of a sheaf morphism graphically, by way of a *commutative diagram*

$$\begin{array}{ccc} \mathscr{S}(a) & \overset{f_a}{\dashrightarrow} & \mathscr{R}(a) \\ {\scriptstyle \mathscr{S}(a\rightsquigarrow b)}\downarrow & & \downarrow{\scriptstyle \mathscr{R}(a\rightsquigarrow b)} \\ \mathscr{S}(b) & \underset{f_b}{\dashrightarrow} & \mathscr{R}(b) \end{array}$$

which we say *commutes*, that is $f_b \circ \mathscr{S}(a \rightsquigarrow b) = \mathscr{R}(a \rightsquigarrow b) \circ f_a$.

Example 3.8 The diagram below exhibits a morphism (dotted arrows) between two sheaves in which the restriction maps are either identities or sums of components

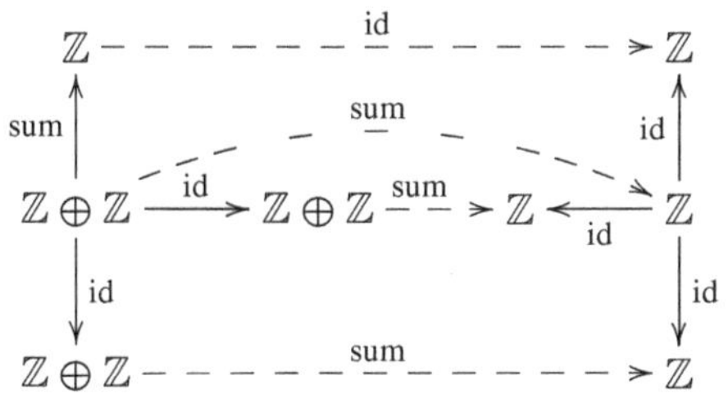

$$\begin{array}{ccccccc} \mathbb{Z} & & & \overset{\text{id}}{\dashrightarrow} & & & \mathbb{Z} \\ {\scriptstyle \text{sum}}\uparrow & & & & & & \uparrow{\scriptstyle \text{id}} \\ \mathbb{Z}\oplus\mathbb{Z} & \overset{\text{id}}{\longrightarrow} & \mathbb{Z}\oplus\mathbb{Z} & \overset{\text{sum}}{\dashrightarrow} & \mathbb{Z} & \underset{\text{id}}{\longleftarrow} & \mathbb{Z} \\ \downarrow{\scriptstyle \text{id}} & & & & & & \downarrow{\scriptstyle \text{id}} \\ \mathbb{Z}\oplus\mathbb{Z} & & & \overset{\text{sum}}{\dashrightarrow} & & & \mathbb{Z} \end{array}$$

(with a dashed arrow labelled sum from the left $\mathbb{Z}\oplus\mathbb{Z}$ to the right middle $\mathbb{Z}$)

Notice that every rectangle in the diagram commutes, for instance

$$\underset{\mathbb{Z}}{\text{id}} \circ \text{sum} = \text{sum} \circ \underset{\mathbb{Z}\oplus\mathbb{Z}}{\text{id}} .$$

Exercise 3.4 Suppose that X and Y are two directed graphs which have the same undirected graph structure (see Fig. 3.6). Show that the sheaves of piecewise linear functions $\mathscr{PL}$ as defined in Example 3.5 constructed over X and Y are isomorphic, by constructing an explicit sheaf isomorphism between them.

Example 3.9 Sampling theory is conveniently studied by using sheaf morphisms. For instance, consider the sheaf of continuous functions defined in Example 3.4 and the $\mathbb{R}$-sampling sheaf defined in Example 3.3. These two sheaves are related; one obtains a discrete timeseries from a continuous signal by sampling. This is encoded

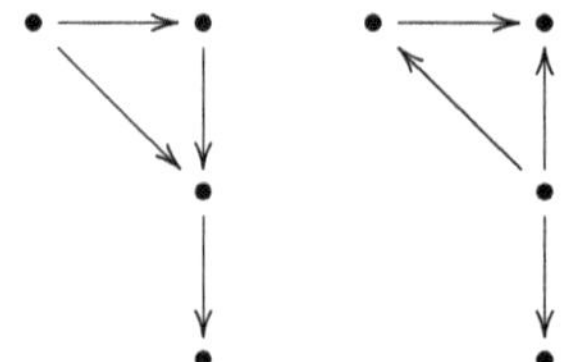

Fig. 3.6 Two directed graphs with the same undirected graph structure

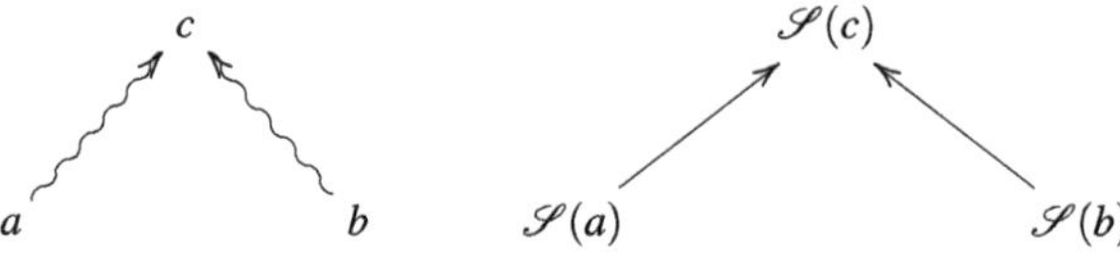

Fig. 3.7 An attachment diagram (*left*) and sheaf (*right*) in which a high dimensional cell (c) is attached to two lower dimensional cells (a and b)

in the construction

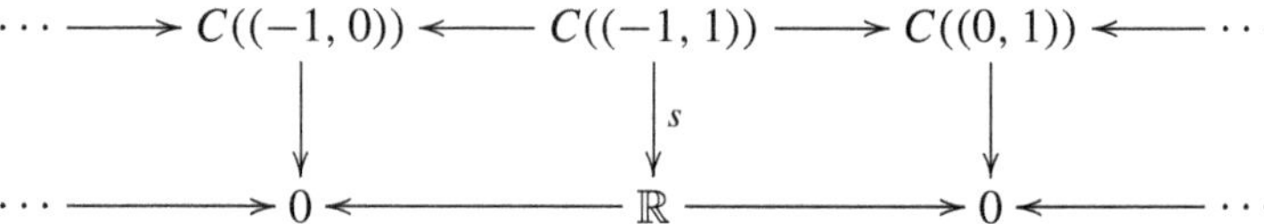

in which the function s takes a continuous function f and returns its value $f(0)$ at 0. Since the diagram commutes (trivially, since half of the maps are zero), this is a sheaf morphism.

3.2 Global Sections

Suppose that $\mathscr{S}$ is a sheaf on a cell complex X. For each cell a of X, an element in $\mathscr{S}(a)$ (a "datum assigned at a") determines elements in $\mathscr{S}(c)$ for each cell c that is attached to a. If a cell c is attached to two lower-dimensional faces $a \rightsquigarrow c$ and $b \rightsquigarrow c$ (see Fig. 3.7), then an element in $\mathscr{S}(a)$ and an element in $\mathscr{S}(b)$ will typically each determine different elements of $\mathscr{S}(c)$. When they happen to determine the same element in $\mathscr{S}(c)$ is a special circumstance, and represents *consistency* between local information.

Definition 3.3 Suppose $\mathscr{S}$ is a sheaf on a cell complex X and that $\mathscr{C}$ and $\mathscr{D}$ are collections of cells of X. An assignment s which assigns an element of $\mathscr{S}(c)$ to each cell $c \in \mathscr{C}$ is called a *section supported on* $\mathscr{C}$ if for each attachment $a \rightsquigarrow b$ of cells in $\mathscr{C}$, $(\mathscr{S}(a \rightsquigarrow b))\ s(a) = s(b)$. Notice that $s(a)$ is an element in the domain of $\mathscr{S}(a \rightsquigarrow b)$ and $s(b)$ is in its range. If r and s are sections supported on $\mathscr{C} \subseteq \mathscr{D}$

respectively, in which $r(a) = s(a)$ for each $a \in \mathscr{C}$, we say that s *extends* r or r *is a restriction of* s. A *global section* is a section supported on all of X.

Example 3.10 Consider the sheaf of sets given by the following diagram

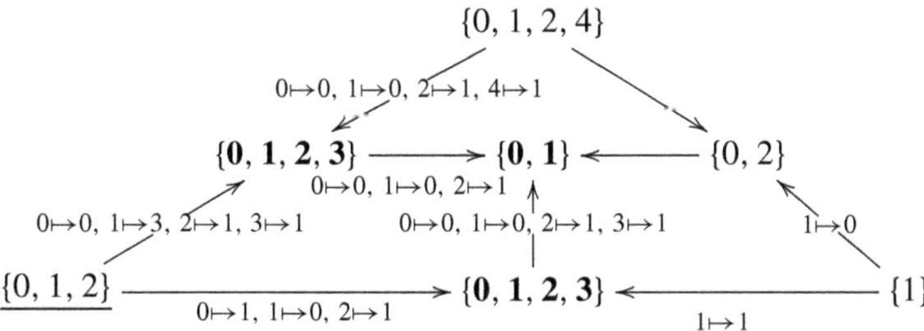

in which the restriction maps are given by explicitly listing their inputs and outputs. The following section supported on the boldfaced cells

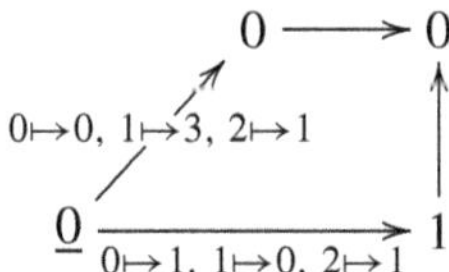

can be extended to the underlined vertex, by assigning a value of 0 there. In contrast, the section

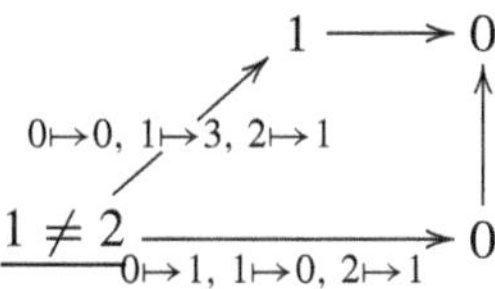

cannot be extended to the underlined vertex, because the restriction maps require it to have value 1 and 2 simultaneously.

The previous example indicates that some local sections of a sheaf of sets may not extend to global sections. The same is true of sheaves of vector spaces. However, if $\mathscr{S}$ is a sheaf of vector spaces, then the collection of sections supported on a given set $\mathscr{C}$ forms a vector space, denoted by $\mathscr{S}(\mathscr{C})$. If Y is a subcomplex of X, we usually abuse notation slightly and write $\mathscr{S}(Y)$.

Example 3.11 Consider the cell complex X shown at left in Fig. 3.8 and the sheaf $\mathscr{V}$ of vector spaces on the right. For the collection $\mathscr{C}_1 = \{v_1, e_1, e_2\}$, the space of sections $\mathscr{V}(\mathscr{C}_1) = \mathbb{R}^2$. The easiest way to see this is to observe that this space of sections is parametrized by the value at the vertex v_1. In the same way, the space of sections over $\mathscr{C}_2 = \{v_2, e_1, e_3\}$ is parametrized by a value at v_2, and so is a 1-dimensional vector space.

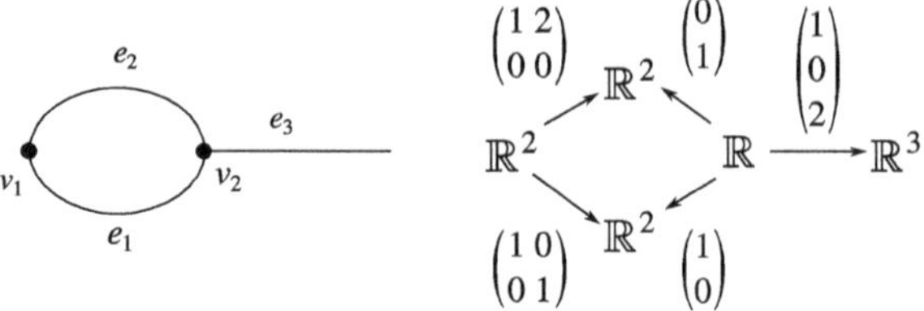

Fig. 3.8 The cell complex X (*left*) and sheaf $\mathscr{V}$ (*right*) in Example 3.11

Enlargening the set of support for the local sections typically reduces the size of the space of sections. For instance, consider $\mathscr{C}_3 = \{v_1, v_2, e_1, e_3\}$, for which the space of sections is $\mathscr{V}(\mathscr{C}_3) = \mathbb{R}$. Notice that any section supported on $\mathscr{C}_3$ specifies values s_1 and s_2 at v_1 and v_2 respectively, which must agree according to

$$\begin{pmatrix}1&0\\0&1\end{pmatrix} s_1 = \begin{pmatrix}1\\0\end{pmatrix} s_2.$$

For this to be satisfied, it is both necessary and sufficient that s_1 is a multiple of $\begin{pmatrix}1\\0\end{pmatrix}$. The space of global sections is smaller yet, since any such section must also satisfy

$$\begin{pmatrix}1&2\\0&0\end{pmatrix} s_1 = \begin{pmatrix}0\\1\end{pmatrix} s_2.$$

The only possibilities are $s_1 = \begin{pmatrix}0\\0\end{pmatrix}$ and $s_2 = 0$. Therefore, $\mathscr{V}(X)$ is the trivial vector space.

Example 3.12 The global sections of the sheaf $\mathscr{PL}$ defined in Example 3.5 are *piecewise linear functions* on a graph G. The stalks of $\mathscr{PL}$ specify the value of the function and the slopes of the function on the edges.

Exercise 3.5 Extend the sheaf of piecewise linear functions $\mathscr{PL}$ to a sheaf of Taylor polynomials on a graph G.

Proposition 3.1 *A morphism between sheaves of vector spaces induces a linear map between spaces of global sections. Isomorphic sheaves have isomorphic spaces of global sections.*

Proof Suppose that $f : \mathscr{R} \to \mathscr{S}$ is morphism of sheaves on a cell complex X. Let r be a global section of $\mathscr{R}$. For each $c \in X$, define $(f(r))\,(c) = f_c\,(r(c))$. Suppose that r, s are global sections of $\mathscr{R}$ and that a, b are scalars. Then for $c \in X$,

$$\begin{aligned}(f(ar+bs))(c) &= f_c((ar+bs)(c))\\ &= f_c(ar(c)+bs(c))\\ &= af_c(r(c))+bf_c(s(c))\\ &= a(f(r))(c)+b(f(s))(c),\end{aligned}$$

so on each stalk, $s \mapsto f(s)$ is linear. Now suppose $a \rightsquigarrow b$, so that

$$\begin{aligned}\mathscr{S}(a \rightsquigarrow b)(f(r))(a) &= \mathscr{S}(a \rightsquigarrow b)f_a(r(a))\\ &= f_b\mathscr{R}(a \rightsquigarrow b)r(a)\\ &= f_b(r(b))\\ &= (f(r))(b),\end{aligned}$$

so $f(r)$ is a section of $\mathscr{S}$.

Now if f is an isomorphism, this means that f_c is an isomorphism for each cell c. If $f(r)$ is zero on all cells, then by definition $f_c(r(c))$ is zero for each cell c. Since f_c is an isomorphism, this means that r is zero on all cells; so $f: \mathscr{R}(X) \to \mathscr{S}(X)$ is injective because of the linearity we just showed. Finally, suppose s is a global section of $\mathscr{S}$. Since f is an isomorphism, for each cell a there is an $r(a)$ with $f_a(r(a)) = s(a)$. If $a \rightsquigarrow b$, then for this assignment r,

$$\begin{aligned}s(b) &= \mathscr{S}(a \rightsquigarrow b)s(a)\\ &= \mathscr{S}(a \rightsquigarrow b)f_a(r(a))\\ &= f_b\mathscr{R}(a \rightsquigarrow b)r(a)\\ &= f_b(r(b))\end{aligned}$$

hence $\mathscr{R}(a \rightsquigarrow b)r(a) = r(b)$, so r is a global section of $\mathscr{R}$. Thus f is therefore surjective. □

Much of the theory of sheaves is concerned with computing spaces of sections and identifying obstructions to extending sections. Chapter 4 defines *cohomology*, the primary tool for computing extensions of sections. However, for simple cases, extensions and induced maps can be computed directly.

Example 3.13 Consider the sheaves

$$\mathscr{S} = \cdots \longrightarrow C((-1,0)) \longleftarrow C((-1,1)) \longrightarrow C((0,1)) \longleftarrow \cdots$$

and

$$\mathscr{T} = \cdots \longrightarrow 0 \longleftarrow \mathbb{R} \longrightarrow 0 \longleftarrow \cdots$$

from Examples 3.4 and 3.3. Both are sheaves over the real line with the usual cellular structure. Observe that many local sections of $\mathscr{S}$ do not extend to global sections, but that all local sections of $\mathscr{T}$ extend to global sections. For instance, the section which assigns $\cos(2\pi x)$ (defined for $x \in (-1, 1)$) to the vertex 0 and $\sin(2\pi x)$ (defined for $x \in (0, 2)$) to the vertex 1 does not extend to the edge $(0, 1)$ since these two functions take different values on the interval $(0, 1)$.

The space of global sections of $\mathscr{S}$ is the space of continuous functions on the real line, so the induced map described in Proposition 3.1 takes a continuous function to a sequence of real numbers. Using the morphism in Example 3.9, this induced map is given by

$$f \mapsto (\ldots, f(-1), f(0), f(1), \ldots).$$

The topology of the base space can strongly impact the global sections of a sheaf. If the base space has nontrivial topology, it often has the effect of preventing sections from being extended.

Example 3.14 Consider the following two sheaves of sets:

$$\cdots \xleftarrow{-1} \mathbb{Z} \xrightarrow{+1} \mathbb{Z} \xleftarrow{-1} \mathbb{Z} \xrightarrow{+1} \mathbb{Z} \xleftarrow{-1} \cdots$$

and

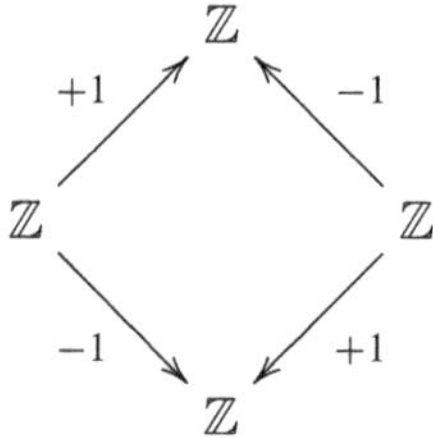

Observe that these two sheaves have the same restrictions and stalks, but different topology. The first has many global sections (they are parametrized by $\mathbb{Z}$). The second has no global sections. For instance, the value of a section in the first sheaf could be any integer n on any of the cells. In the second sheaf, there is no consistent choice of values on all of the cells.

3.3 Operations on Sheaves

Sheaves are especially convenient for manipulating sets of local measurements because they have a number of natural operations that preserve their structure.

3.3.1 Pushforwards and Pullbacks

The definition of a sheaf morphism can be extended to handle sheaves on different spaces. This plays an important role when relating information collected in different contexts. It relies on the notion of spaces of local sections as follows.

Definition 3.4 A *morphism* $m : \mathscr{S} \to \mathscr{R}$ from a sheaf $\mathscr{S}$ over a space Y to a sheaf $\mathscr{R}$ on X consists of a cellular map $f : X \to Y$ and a collection of linear maps $m_x : \mathscr{S}(f(x)) \to \mathscr{R}(x)$ such that for each $x \rightsquigarrow y$ in X, the following diagram commutes

$$\begin{array}{ccc} \mathscr{S}(f(x)) & \xrightarrow{m_x} & \mathscr{R}(x) \\ {\scriptstyle \mathscr{S}(f(x)\rightsquigarrow f(y))}\downarrow & & \downarrow{\scriptstyle \mathscr{R}(x\rightsquigarrow y)} \\ \mathscr{S}(f(y)) & \xrightarrow{m_y} & \mathscr{R}(y) \end{array}$$

Sometimes we will say that m is a sheaf morphism *along* f.

Example 3.15 Rainfall and snowfall are typically measured at discrete locations using rain or snow gauges. Suppose that the rainfall and snowfall over a region X are represented by a continuous function $f : X \to \mathbb{R}^2$. We can encode this as the sheaf $\mathscr{P}$ over a space with one vertex $*$ whose stalk is $C(X, \mathbb{R}^2)$, the set of continuous $\mathbb{R}^2$-valued functions.

A commonly-used (though faulty) heuristic is that a snowfall of 10 cm corresponds to the same amount of precipitation as 1 cm of rain. Using this heuristic, the rain-equivalent amount of precipitation at a location $x \in X$ is given by the product

$$\begin{pmatrix} 1 & 10 \end{pmatrix} f(x).$$

If there are N rain gauges at locations $x_1, \ldots, x_N$ and M snow gauges at locations $y_1, \ldots, y_M$, we can represent the collection of rain-equivalent measurements as a sheaf $\mathscr{M}$ over the discrete set of gauge locations $G = \{x_1, \ldots, x_N, y_1, \ldots, y_M\}$. This sheaf simply consists of the assignment of $\mathbb{R}$ to each point in G.

The process of making precipitation measurements corresponds to a sheaf morphism $m : \mathscr{P} \to \mathscr{M}$, given by the diagram

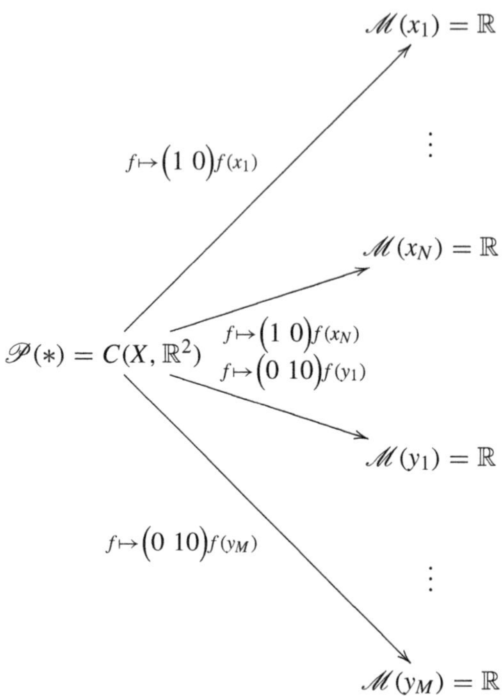

Observe that the space of global sections of $\mathscr{P}$ is given by its single stalk, namely $C(X, \mathbb{R}^2)$. On the other hand, the space of global sections of $\mathscr{M}$ is $\mathbb{R}^{M+N}$. A straightforward calculation shows that the sheaf morphism m induces a linear map that takes $f \in C(X, \mathbb{R}^2)$ to

$$\begin{pmatrix} (1\ 0) f(x_1) \\ \vdots \\ (1\ 0) f(x_N) \\ (0\ 10) f(y_1) \\ \vdots \\ (0\ 10) f(y_M) \end{pmatrix}.$$

This example is an instance of a *sampling morphism* (see also Example 3.9), which is treated more extensively in Sect. 4.5.

Remark 3.3 Some authors, notably Bredon (1997), call a general sheaf morphism an "f-cohomomorphism." We will not use this notation as it seems to complicate matters.

The following example indicates how sections and morphisms are related.

Example 3.16 Suppose that $f : X \to *$ is the map which collapses a cell complex X to a single point. Each global section s of a sheaf $\mathscr{S}$ over X defines a morphism from the (unique) sheaf over a single point $*$ whose stalk is the field with two elements $\mathbb{F}_2$, because then we simply define $m_x = s(x)$. The axioms for global sections given in Definition 3.3 are precisely what is needed to show that m is a morphism.

Definition 3.5 Suppose that $f : X \to Y$ is a cellular map and that $\mathscr{S}$ is a sheaf on Y. The *pullback* $f^*\mathscr{S}$ is a sheaf on X given by

$$(f^*\mathscr{S})(c) = \mathscr{S}(f(c)),$$

and

$$(f^*\mathscr{S})(a \rightsquigarrow b) = \mathscr{S}(f(a) \rightsquigarrow f(b)),$$

which is an identity map if $f(a) = f(b)$.

Lemma 3.1 *If $f : X \to Y$ is a cellular map and $\mathscr{S}$ is a sheaf on Y, then $f^* : \mathscr{S} \to f^*\mathscr{S}$ is a sheaf morphism given by the collection of maps $f_x^* : \mathscr{S}(f(x)) \to f^*\mathscr{S}(x)$ which satisfy the equation*

$$f_x^* s = s(f(x))$$

for each cell x in X and $s \in \mathscr{S}(f(x))$.

Proof We will compute both branches of the commutative diagram in Definition 3.4. Suppose that $x \rightsquigarrow y$ in X and that $s \in \mathscr{S}(f(x))$.

1. Lower branch:

$$\begin{aligned}
\left(\left(f_y^* \circ f^*\mathscr{S}(x \rightsquigarrow y)\right) s\right)(y) &= \left(\left(f_y^* \circ \mathscr{S}(f(x) \rightsquigarrow f(y))\right) s\right)(y) \\
&= \left(f_y^*(\mathscr{S}(f(x) \rightsquigarrow f(y))s)\right)(y) \\
&= (\mathscr{S}(f(x) \rightsquigarrow f(y))s)(f(y)) \\
&= \mathscr{S}(f(x) \rightsquigarrow f(y))s(f(x))
\end{aligned}$$

2. Upper branch:

$$\begin{aligned}
(f^*\mathscr{S})(x \rightsquigarrow y)f_x^* s &= (f^*\mathscr{S})(x \rightsquigarrow y)s(f(x)) \\
&= \mathscr{S}(f(x) \rightsquigarrow f(y))s(f(x))
\end{aligned}$$

□

Example 3.17 Figure 3.9 shows an example of a cellular map $f : X \to Y$ and a sheaf $\mathscr{R}$ on Y. The pullback sheaf $f^*\mathscr{R}$ has the same stalks and the same restrictions as $\mathscr{R}$. The pullback morphism $f^* : \mathscr{R} \to f^*\mathscr{R}$ is indicated by the right-to-left dashed arrows in the figure.

Definition 3.6 Suppose x is a cell of a cell complex X. The *star* of x is the set

$$\operatorname{star} x = \{x\} \cup \{y : x \rightsquigarrow y\}.$$

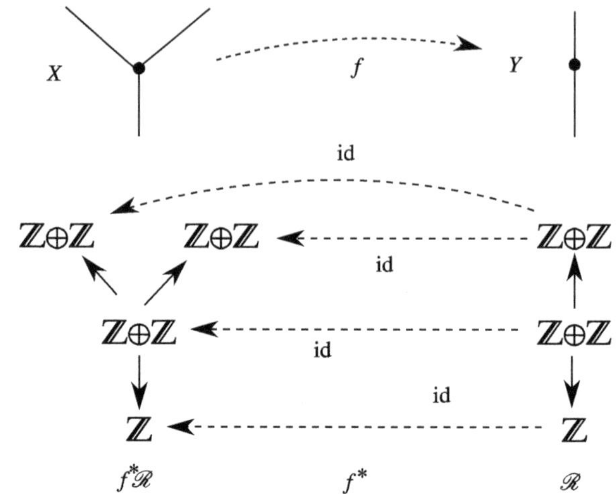

Fig. 3.9 Pullback of a sheaf $\mathscr{R}$ along a cellular map $f : X \to Y$. Note in particular that the maps which constitute the sheaf morphism travel in the opposite direction as the base space map

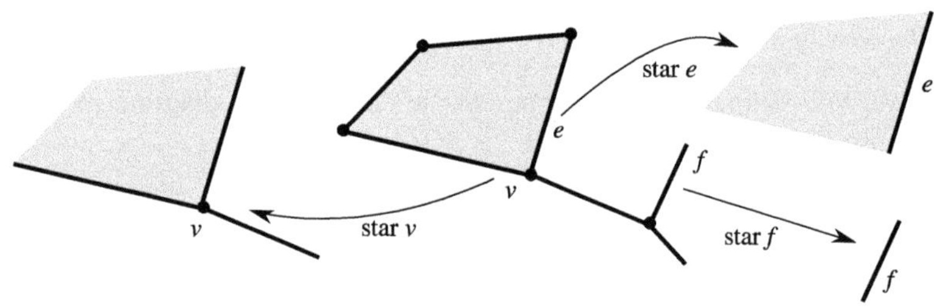

Fig. 3.10 The star of several cells in a cell complex

Typically, as Fig. 3.10 shows, the star of a cell in a cell complex is *not* itself a cell complex. Caution should therefore be exercised when computing with stars.

Definition 3.7 Suppose $f : X \to Y$ is a cellular map and that $\mathscr{R}$ is a sheaf on X. The *pushforward* $f_*\mathscr{R}$ is a sheaf on Y given by

$$(f_*\mathscr{R})(c) = \mathscr{R}(f^{-1}(\text{star } c)),$$

in which the restriction maps $(f_*\mathscr{R})(a \rightsquigarrow b)$ are given by restricting a section s over $f^{-1}(\text{star } a)$ to one over $f^{-1}(\text{star } b)$.

Lemma 3.2 *If $f : X \to Y$ is a cellular map and $\mathscr{R}$ is a sheaf on X, then $f_* : f_*\mathscr{R} \to \mathscr{R}$ is a sheaf morphism given by*

$$(f_*)_x s = s(x)$$

where $x \in X$, and $s \in f_*\mathscr{R}(f(x)) = \mathscr{R}(f^{-1}(\text{star } f(x)))$.

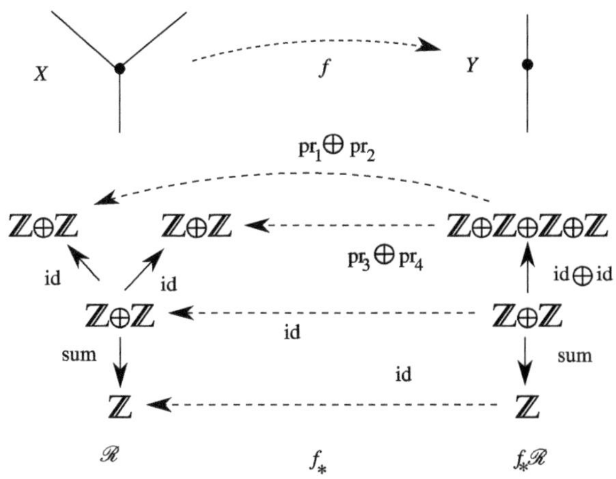

Fig. 3.11 The pushforward of a sheaf $\mathscr{R}$

Proof We will again compute both branches of the commutative diagram in Definition 3.4. Suppose that $x \rightsquigarrow y$ in X and that $s \in f_*\mathscr{R}(f(x))$.

1. Lower branch:

$$\left(\left((f_*)_y \circ f_*\mathscr{R}(f(x) \rightsquigarrow f(y))\right) s\right)(y) = (f_*\mathscr{R}(f(x) \rightsquigarrow f(y))s)(y)$$
$$= (\mathscr{R}(x \rightsquigarrow y)s(x))(y)$$

2. Upper branch:

$$(\mathscr{R}(x \rightsquigarrow y)(f_*)_x s)(y) = (\mathscr{R}(x \rightsquigarrow y)s(x))(y)$$

□

Again notice that the direction of the sheaf morphism $f_* : f_*\mathscr{R} \to \mathscr{R}$ is opposite from the base space map $f : X \to Y$.

Remark 3.4 Because computing the pushforward involves computing the space of global sections over the star of a cell, pushforwards are much more difficult to manipulate in practice than pullbacks.

Example 3.18 Figure 3.11 shows an example of a cellular map $f : X \to Y$ and a sheaf $\mathscr{R}$ on X. The pushforward sheaf $f_*\mathscr{R}$ has stalks that correspond to spaces of global sections of $\mathscr{R}$. From a computational standpoint, this means that the stalks of $f_*\mathscr{R}$ are direct sums of stalks of $\mathscr{R}$. The pushforward morphism $f_* : f_*\mathscr{R} \to \mathscr{R}$ is indicated by the right-to-left dashed arrows in the figure.

Pushforwards and pullbacks are important operations because they give rise to a canonical way to address the effects of "switching" base spaces and the effects of morphisms directly on the stalks.

Proposition 3.2 *Every morphism $m\colon \mathscr{S} \to \mathscr{R}$ uniquely factors in two ways, such that the following diagram of sheaf morphisms commutes*

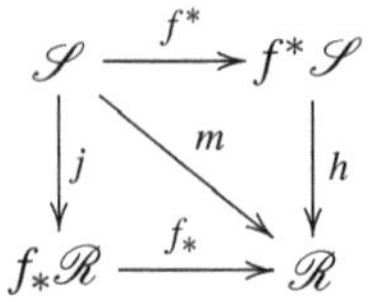

$$\begin{array}{ccc} \mathscr{S} & \xrightarrow{f^*} & f^*\mathscr{S} \\ {\scriptstyle j}\downarrow & \overset{m}{\searrow} & \downarrow{\scriptstyle h} \\ f_*\mathscr{R} & \xrightarrow{f_*} & \mathscr{R} \end{array}$$

where h is a morphism of sheaves on X and j is a morphism of sheaves on Y.

Proof (Following Bredon (1997, Sect. I.4)) Let us consider the upper branch of the diagram, and define the morphism h of sheaves on X by

$$h_x = m_x \circ \left(f_x^*\right)^{-1}.$$

Evidently, there is exactly one such choice for h_x. Observe that $\left(f_x^*\right)^{-1}$ is well-defined because $f_x^* : \mathscr{S}(f(x)) \to f^*\mathscr{S}(x)$ is an isomorphism of vector spaces wherever it is defined. Hence by the Lemma

$$f_y^* \circ \mathscr{S}(f(x) \rightsquigarrow f(y)) = \mathscr{S}(f(x) \rightsquigarrow f(y)) \circ f_x^*,$$

by multiplying both sides of the equation by inverses, we obtain

$$\mathscr{S}(f(x) \rightsquigarrow f(y)) \circ \left(f_x^*\right)^{-1} = \left(f_y^*\right)^{-1} \circ \mathscr{S}(f(x) \rightsquigarrow f(y)),$$

so that

$$\begin{aligned} h_y \circ f^*\mathscr{S}(x \rightsquigarrow y) &= m_y \circ \left(f_y^*\right)^{-1} \circ \mathscr{S}(f(x) \rightsquigarrow f(y) \\ &= m_y \circ \mathscr{S}((f(x) \rightsquigarrow f(y)) \circ \left(f_x^*\right)^{-1} \\ &= \mathscr{R}(x \rightsquigarrow y) \circ m_x \circ \left(f_x^*\right)^{-1} \\ &= \mathscr{R}(x \rightsquigarrow y) \circ h_x \end{aligned}$$

so that h is a morphism of sheaves.

The pushforward is a bit more subtle. For a cell $y \in Y$, the definition of the morphism m implies that there are maps $m_y : \mathscr{S}(y) \to \mathscr{R}(z)$ for each $z \in f^{-1}(y)$. However, since these maps come from a morphism, they commute with the restrictions. Therefore, each *value* in the stalk $\mathscr{S}(y)$ determines a *section* over the star of $f^{-1}(y)$ in $\mathscr{R}$. But this is precisely the definition of $f_*\mathscr{R}(y)$. Hence, we have defined a morphism $j\colon \mathscr{S} \to f_*\mathscr{R}$. □

This Proposition is helpful because it implies pushforwards and pullbacks are *functors* (Definition 4.2), a concept which plays an essential role in the later chapters of this book.

Proposition 3.3 *Suppose that $f : X \to Y$ is a cellular map and that $m : \mathscr{R} \to \mathscr{S}$ is a morphism of sheaves on X. Then there is a unique morphism $f_* m$ that makes the diagram below commute*

$$\begin{array}{ccc} f_*\mathscr{R} & \xrightarrow{f_*} & \mathscr{R} \\ \downarrow{\scriptstyle f_*m} & & \downarrow{\scriptstyle m} \\ f_*\mathscr{S} & \xrightarrow{f_*} & \mathscr{S} \end{array}$$

When $\mathscr{R}$ and $\mathscr{S}$ are sheaves on Y, then there is a unique morphism f^*m that makes

$$\begin{array}{ccc} \mathscr{R} & \xrightarrow{f^*} & f^*\mathscr{R} \\ \downarrow{\scriptstyle m} & & \downarrow{\scriptstyle f^*m} \\ \mathscr{S} & \xrightarrow{f^*} & f^*\mathscr{S} \end{array}$$

commute.

Proof The trick is to notice that by Proposition 3.2 the diagram includes a diagonal morphism j

$$\begin{array}{ccc} f_*\mathscr{R} & \xrightarrow{f_*} & \mathscr{R} \\ \downarrow{\scriptstyle f_*m} & \searrow^{j} & \downarrow{\scriptstyle m} \\ f_*\mathscr{S} & \xrightarrow{f_*} & \mathscr{S} \end{array}$$

The same argument as was used in the proof of Proposition 3.2 can be used to define f_*m. Therefore the diagram commutes.

For the pullback, we merely need to define $(f^*m)_{f(x)} = f^*_{f(x)} \circ m_{f(x)} \circ (f^*_{f(x)})^{-1}$ since the pullback morphism f^*_y is an isomorphism on stalks. (Notice that the left $f^*_{f(x)}$ in the definition of f^*m is on $\mathscr{S}$ and the right $f^*_{f(x)}$ is on $\mathscr{R}$). □

3.3.2 Algebraic Operations

Definition 3.8 Given two sheaves of vector spaces $\mathscr{R}$ and $\mathscr{S}$ on a cell complex X, their *sum* $\mathscr{R} \oplus \mathscr{S}$ is given by

$$(\mathscr{R} \oplus \mathscr{S})(c) = \mathscr{R}(c) \oplus \mathscr{S}(c),$$

and

$$(\mathscr{R} \oplus \mathscr{S})(c \rightsquigarrow d)(v, w) = (\mathscr{R}(c \rightsquigarrow d)v, \mathscr{S}(c \rightsquigarrow d)w),$$

for $v \in \mathscr{R}(c)$ and $w \in \mathscr{S}(c)$. We can proceed in the same way to define $\mathscr{R} \otimes \mathscr{S}$, the tensor product of two sheaves.

Example 3.19 Flow sheaves are useful for representing the transport of individual commodities along a network. The sum of two flow sheaves therefore represents the transport of two independent commodities on the same network. Suppose that apples and oranges are being transported on a network N. Let the flow sheaf $\mathscr{A}$ represent the amount of apples being transported on N. For instance, $\mathscr{A}(e)$ is the total number of apples being transported along an edge e. Similarly, let $\mathscr{O}$ represent the number of oranges in transit. The sum $\mathscr{A} \oplus \mathscr{O}$ represents the numbers of both kinds of fruit on the network. Therefore $(\mathscr{A} \oplus \mathscr{O})(e) = \mathbb{N}^2$, representing the number of apples and oranges on the edge e.

Definition 3.9 Suppose $\mathscr{R}$ and $\mathscr{S}$ are sheaves on X, and that for each cell c, $\mathscr{R}(c) \subseteq \mathscr{S}(c)$ and $\mathscr{S}(a \rightsquigarrow b)$ restricted to $\mathscr{R}(a)$ is equal to $\mathscr{R}(a \rightsquigarrow b)$. We call $\mathscr{R}$ a *subsheaf* of $\mathscr{S}$ in this case. We form the *quotient sheaf* $\mathscr{S}/\mathscr{R}$ by defining

$$(\mathscr{S}/\mathscr{R})(c) = \mathscr{S}(c)/\mathscr{R}(c),$$

and

$$(\mathscr{S}/\mathscr{R})(a \rightsquigarrow b)v = \left(\pi_c \circ \mathscr{S}(a \rightsquigarrow b) \circ \pi_c^{-1}\right)(v),$$

where $\pi_c : \mathscr{S}(c) \to \mathscr{S}(c)/\mathscr{R}(c)$ is the canonical projection. (Observe that since $\mathscr{S}(a \rightsquigarrow b)$ is linear, the definition is well-defined.)

The definition of a quotient for sheaves of groups also makes sense if $\mathscr{R}(c)$ is a normal subgroup of $\mathscr{S}(c)$.

Example 3.20 As a simple example of a subsheaf, consider the sheaf $\mathscr{S}$ (center, below). A subsheaf $\mathscr{R}$ of $\mathscr{S}$ is shown at left below, where its inclusion into $\mathscr{S}$ is shown with dotted arrows. Their quotient $\mathscr{S}/\mathscr{R}$ is shown at right, below.

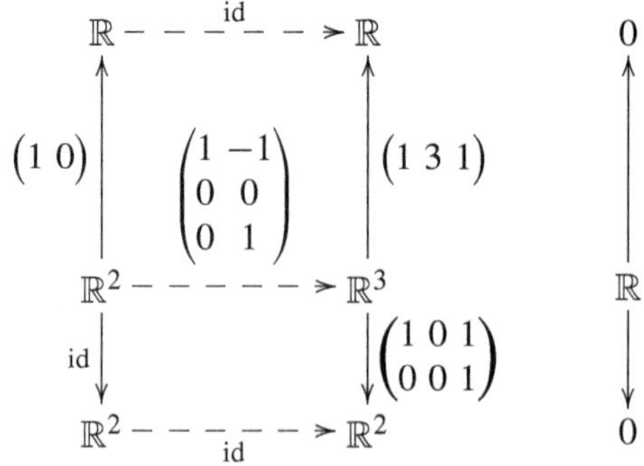

Quotient sheaves will play an important role in sampling theory (Sect. 4.5), though they also arise in much simpler settings. For instance, consider the sheaf $\mathscr{PL}$ of piecewise linear functions (defined in Example 3.5) on a graph X. This sheaf has a constant $\mathscr{K}$ as a subsheaf, in which the stalk at a vertex $c \in X$ is given by $\mathscr{K}(c) = \{(y, 0, \ldots, 0) : y \in \mathbb{R}\}$. Observe that the restriction in $\mathscr{PL}$ between two cells $c \rightsquigarrow d$

$$\mathscr{PL}(c \rightsquigarrow d)(y, 0, \ldots, 0) = (y, 0),$$

induces the identity map $\mathscr{K}(c \rightsquigarrow d) = \text{id}$ as a restriction in the sheaf $\mathscr{K}$. The quotient $\mathscr{PL}/\mathscr{K}$ represents all piecewise linear functions defined up to the addition of a constant. All of the nontrivial elements of $\mathscr{PL}/\mathscr{K}$ consist of non-constant functions.

3.4 Case Study: Topological Filters

Filters are the backbone of signal processing. In this case study, we show that sheaf operations enable the description of sophisticated, possibly nonlinear filters from simpler, locally-defined ones. In particular, a pair of sheaf morphisms can be used to describe any finite impulse response filter. We show by example that these *topological filters* are the correct way to describe any filter that operates locally on an input signal to produce an output signal. This more flexible definition paves the way for the invention of powerful filters that can robustly highlight features of interest. Specifically, we show that linear shift-invariant filters, the graph Laplacian, constant false alarm detectors, and an angle-valued image filter are all examples of topological filters. All of these filters involve processing adjacent blocks of measurements, which is a convenient (but not strictly necessary) way to construct topological filters.

3.4.1 Linear Shift-Invariant Systems

Suppose that $\{x_n\}_{n=-\infty}^{\infty}$ is a timeseries in a vector space V. A *finite impulse response filter* F of order N is a linear shift-invariant map between sequences in V that can be written in the form

$$y_n = (F(x))_n = \sum_{i=0}^{N} a_i x_{n-i} = L(x_{n-N}, \ldots, x_n), \tag{3.2}$$

where $L : V^{N+1} \rightarrow V$ is linear. This filter produces a sequence $\{y_n\}$ by taking a weighted average of a sliding window of $N + 1$ samples $x_{n-N}, \ldots, x_n$, in which the weights are $\{a_i\}_{i=0}^{N}$.

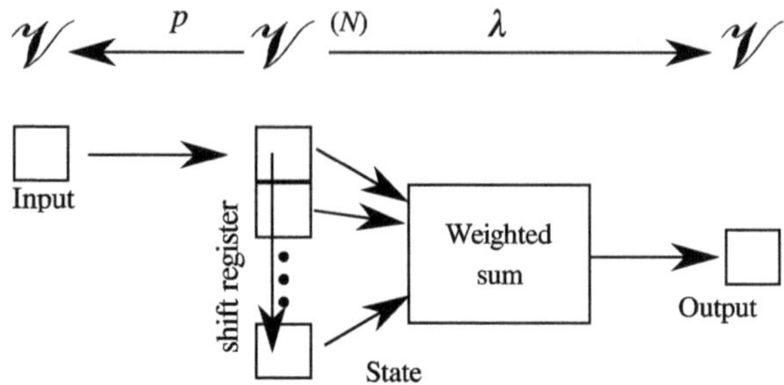

Fig. 3.12 The construction of a linear shift-invariant filter as a sequence of two morphisms between three sheaves

Example 3.21 The *boxcar filter* is a finite impulse response filter in which the weights are all the same. Typically, one chooses $a_i = a = 1/(N+1)$ or $a_i = a = 1/\sqrt{N+1}$ so that the signal's 1-norm or its 2-norm (respectively) are preserved. This filter is a *low pass* filter in that high frequency variations in $\{x_n\}$ are reduced in $\{y_n\}$. The choice of N sets the size of the pass band.

To analyze the frequency response of this filter, one usually makes use of what is called the *z-transform*, which is merely a particular complex power series representation. Suppose that

$$x_n = z^n$$

for some $z \in \mathbb{C}$. For instance, if $z = e^{i\omega}$, then $\{x_n\}$ is a signal with angular frequency[1] ω. Substituting this choice of x_n into (3.2), we obtain

$$\begin{aligned} y_n &= \sum_{i=0}^{N} a_i x_{n-i} = a \sum_{i=0}^{N} z^{n-i} = az^n \sum_{i=0}^{N} z^{-i} \\ &= az^n \frac{z^{-N-1} - 1}{z^{-1} - 1} = \left(a \frac{1 - z^{N+1}}{z^N - z^{N+1}} \right) z^n \\ &= \left(a \frac{1 - z^{N+1}}{1 - z} \right) z^{n-N} = \left(a \frac{1 - z^{N+1}}{1 - z} \right) x_{n-N} \end{aligned}$$

Therefore, the output signal is amplified by a factor of $\left(a \frac{1 - z^{N+1}}{1 - z} \right)$. In particular, whenever $z^{N+1} = 1$ (and $z \neq 1$), the output does not respond at all to the input. The smallest such choice of z sets the bandwidth of the filter, which is $\frac{2\pi}{N+1}$. Therefore, a larger averaging window leads to a filter with a smaller bandwidth.

The construction of a topological filter from a linear shift-invariant one is summarized in Fig. 3.12 and closely parallels the typical hardware implementation of such a filter. We will represent the input signal $\{x_n\}$ as a global section of the

[1] If the period $2\pi/\omega$ is an integer, then $\{x_n\}$ is periodic.

V-sampling sheaf supported on vertices $\mathbb{Z} \subset \mathbb{R}$ for the usual cell complex structure on $\mathbb{R}$ (Example 3.3). Explicitly, the space of global sections of the sheaf $\mathscr{V}$

$$\cdots \longrightarrow 0 \longleftarrow V \longrightarrow 0 \longleftarrow V \longrightarrow \cdots$$

is isomorphic to the vector space of infinite sequences $\mathbb{R}^{\mathbb{Z}}$ in V.

We represent the internal state of the filter with the *M-term grouping sheaf*, $\mathscr{V}^{(M)}$, in which $\mathscr{V} = \mathscr{V}^{(1)}$. This sheaf $\mathscr{V}^{(M)}$ is given by the diagram

$$\cdots \xleftarrow{\sigma_-} V^M \xrightarrow{\sigma_+} V^{M-1} \xleftarrow{\sigma_-} V^M \xrightarrow{\sigma_+} \cdots$$

in which

$$\sigma_-(x_1, \ldots, x_M) = (x_1, \ldots, x_{M-1}) \text{ and } \sigma_+(x_1, \ldots, x_M) = (x_2, \ldots, x_M).$$

The first of the two sheaf morphisms $p\colon \mathscr{V}^{(N+1)} \to \mathscr{V}$, shows how terms of the input $\{x_i\}$ are grouped into adjacent blocks of length $N+1$ for processing in F, stored internal to the filter. We encode this by the diagram

$$\begin{array}{ccccccccc}
\cdots \longrightarrow & V^N & \longleftarrow & V^{N+1} & \longrightarrow & V^N & \longleftarrow & V^{N+1} & \longrightarrow \cdots \\
& \downarrow{\scriptstyle 0} & & \downarrow{\scriptstyle \mathrm{pr}_{N+1}} & & \downarrow{\scriptstyle 0} & & \downarrow{\scriptstyle \mathrm{pr}_{N+1}} & \\
\cdots \longrightarrow & 0 & \longleftarrow & V & \longrightarrow & 0 & \longleftarrow & V & \longrightarrow \cdots
\end{array}$$

in which the top row is $\mathscr{V}^{(N+1)}$, and

$$\mathrm{pr}_{N+1}(x_1, \ldots, x_{N+1}) = x_{N+1} \tag{3.3}$$

is the projection onto the last component.

Exercise 3.6 Show that p induces an isomorphism $P : \mathscr{V}^{(N+1)}(\mathbb{R}) \to \mathscr{V}(\mathbb{R})$ on global sections, thereby showing that all of the $\mathscr{V}^{(N)}$ have isomorphic spaces of global sections.

The second morphism $\lambda\colon \mathscr{V}^{(N+1)} \to \mathscr{V}$ encodes the action of the map L on each block of terms of $\{x_n\}$. Again, this is efficiently defined using the diagram

$$\begin{array}{ccccccccc}
\cdots \longrightarrow & V^N & \longleftarrow & V^{N+1} & \longrightarrow & V^N & \longleftarrow & V^{N+1} & \longrightarrow \cdots \\
& \downarrow{\scriptstyle 0} & & \downarrow{\scriptstyle L} & & \downarrow{\scriptstyle 0} & & \downarrow{\scriptstyle L} & \\
\cdots \longrightarrow & 0 & \longleftarrow & V & \longrightarrow & 0 & \longleftarrow & V & \longrightarrow \cdots
\end{array}$$

This too induces a linear map $\Lambda : \mathscr{V}^{(N+1)}(\mathbb{R}) \to \mathscr{V}(\mathbb{R})$ on the space of global sections, though not an isomorphism. Constructing the two morphisms one after another

$$\mathscr{V} \xleftarrow{p} \mathscr{V}^{(N+1)} \xrightarrow{\lambda} \mathscr{V}$$

results in a particular encoding of a finite impulse response filter.

Example 3.22 Suppose that the order $N = 2$ finite impulse response filter with weights $a_0 = 1/5$, $a_1 = 3/5$, $a_2 = 1/5$ is applied to the data stream $x = \dots, 1, 3, 5, 2, 4, \dots$. This results in the following diagram of sections of the sheaves:

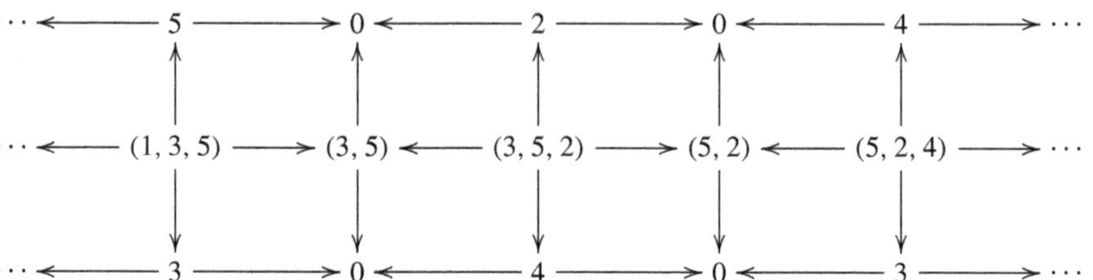

Observe that the middle row specifies which data is being stored in order to compute each output value, and the output data stream $\dots, 3, 4, 3, \dots$ can be read directly from the bottom row.

Proposition 3.4 *The composition of the two induced maps on global sections represents the linear shift-invariant filter F in the following way*

$$F = \Lambda \circ P^{-1}.$$

Remark 3.5 It is straightforward to generalize the construction to handle infinite impulse response filters, by merely defining $\mathscr{V}^{(\infty)}$ to be the appropriate limit construction of the $\mathscr{V}^{(N)}$.

Proof (of Proposition 3.4) Suppose that $\{x_n\}$ is the sequence we input to F. In that case, we have a section s of $\mathscr{V}$ that is given by $s(n) = x_n$ on each vertex n. The corresponding section $P^{-1}s$ over $\mathscr{V}^{(N+1)}$ at the vertex n will have value $(x_{n-N}, \dots, x_n)$. Finally, the section $\Lambda \circ P^{-1}s$ over n will have value $L(x_{n-N}, \dots, x_n) = (F(x))_n$. □

This indicates that a general diagram of sheaf morphisms

$$\mathscr{S}_1 \xleftarrow{m_1} \mathscr{S}_2 \xrightarrow{m_2} \mathscr{S}_3$$

should correspond to a kind of *topological filter*. We delay the precise definition until Definition 4.15, though for the rest of this chapter we merely require that m_1 induces an isomorphism on the spaces of global sections $\mathscr{S}_1(X) \to \mathscr{S}_2(X)$.

Example 3.23 The delay filter $(F(x))_n = x_{n-1}$ can be encoded by the diagram

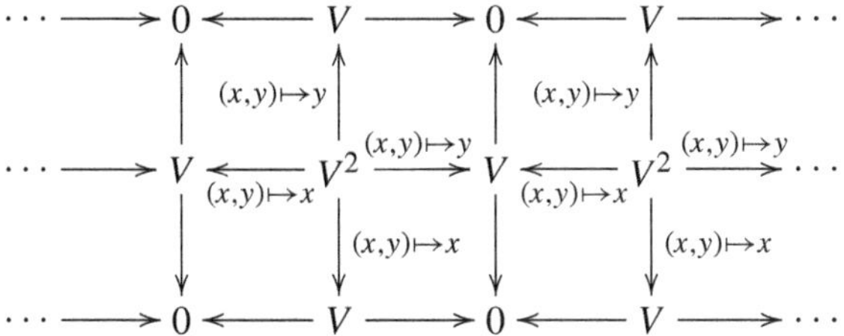

Example 3.24 Consider the filter whose impulse response is zero except for three consecutive terms, all equal to $1/3$. If this filter is presented with the input sequence $\ldots, 1, 1, 9, 2, \ldots$, it will produce the output sequence $\ldots, 2.7, 2.3, 3.7, 4, \ldots$. The encoding described by Proposition 3.4 can be organized into the diagram

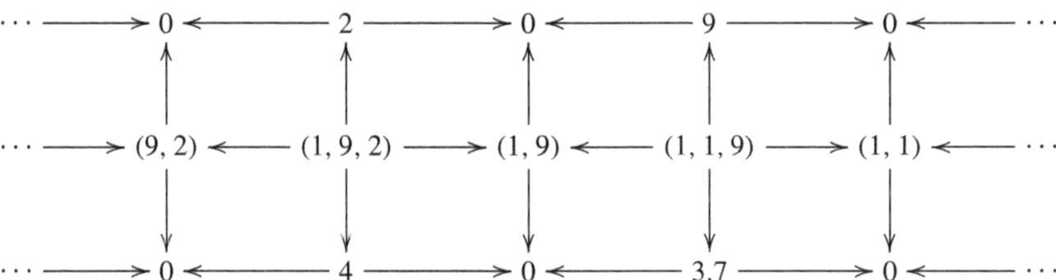

3.4.2 Linear Filtering on Nontrivial Base Spaces

Topological filters encompass certain *non*-shift-invariant filters, such as low pass filters on graphs. This gives considerable expressive power and mirrors the typical implementation of these filters.

We begin by using a generalization of the grouping sheaf employed in Sect. 3.4.1 to graphs. In this example, we show how to generalize $\mathscr{V}^{(3)}$ and obtain a low-pass filter based on the *graph Laplacian*.

Definition 3.10 Given a graph G with N vertices labeled $\{v_1, \ldots, v_N\}$, the *graph Laplacian matrix* L is the $N \times N$ matrix given by

$$L_{i,j} = \begin{cases} \deg v_i \text{ if } i = j \\ -1 \text{ if } v_i \text{ is adjacent to } v_j \\ 0 \text{ otherwise} \end{cases}$$

If f is a real-valued function on the vertices of the graph, then f can be represented by a vector in $\mathbb{R}^N$, in which the i-th component represents the value of f at v_i. Then L can be interpreted as a linear operator on the space of functions on the graph.

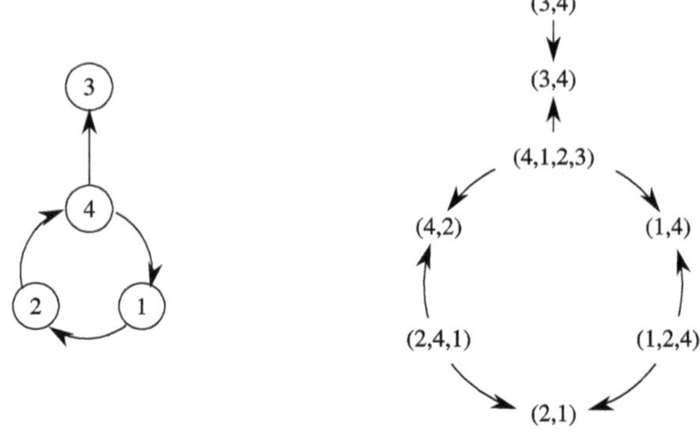

Fig. 3.13 An example of a function on a directed graph (*left*) and its section of the grouping sheaf $\mathscr{V}^{(3)}$ (right)

To realize the operator L as a topological filter, let $\mathscr{V}$ be the $\mathbb{R}$-sampling sheaf on G. That is, let $\mathscr{V}$ assign the stalk $\mathbb{R}$ to each vertex and the trivial vector space to each edge. The space of global sections of $\mathscr{V}$ is $\mathbb{R}^N$.

We construct $\mathscr{V}^{(3)}$ so that the stalk over each edge is $\mathbb{R}^2$ and the stalk over each vertex v_i is $\mathbb{R}^{1+\deg v_i}$. To facilitate the construction, suppose that the edges of G are oriented arbitrarily. Suppose that e is the k-th edge attached to a vertex v. The restriction from $\mathscr{V}^{(3)}(v \rightsquigarrow e)$ is given by its value on a vector $\left(a_1, \ldots, a_{1+\deg v}\right)$

$$\left(\mathscr{V}^{(3)}(v \rightsquigarrow e)\right)\left(a_1, \ldots, a_{1+\deg v}\right) = \begin{cases} (a_1, a_k) & \text{if } e \text{ points into } v \\ (a_k, a_1) & \text{if } e \text{ points away from } v \end{cases}$$

See Fig. 3.13 for an example of $\mathscr{V}^{(3)}$ on a graph. (This generalizes the construction for $\mathscr{V}^{(3)}$ given in the previous section since the vertices in the usual cell complex structure for $\mathbb{R}$ all have degree 2.)

Exercise 3.7 Show that the linear map $P: \mathscr{V}^{(3)}(G) \to \mathscr{V}(G)$ induced on the space of global sections by the sheaf morphism p is an isomorphism of vector spaces. Hint: do Exercise 3.6 first.

Exercise 3.8 Extend the construction of $\mathscr{V}^{(3)}$ to construct generalizations of $\mathscr{V}^{(N)}$ on graphs. Hint: what should the stalk dimension over an edge be?

The construction of the topological filter proceeds as before, using the diagram of sheaves

$$\mathscr{V} \xleftarrow{p} \mathscr{V}^{(3)} \xrightarrow{\lambda} \mathscr{V}.$$

We construct the morphisms p and λ by their component maps as before. The morphism p must be the zero map on each edge, but it is the projection $\mathbb{R}^{1+\deg v} \to \mathbb{R}$ onto the first component for each vertex. This too generalizes the construction of the morphism p for shift-invariant filters.

The morphism λ is defined to capture the local structure of the graph Laplacian matrix, namely its component maps have the form

$$\lambda_v\left(a_1, \ldots, a_{1+\deg v}\right) = a_1 \deg v - \sum_{i=2}^{1+\deg v} a_i.$$

Proposition 3.5 *The graph Laplacian $L = \Lambda \circ P^{-1}$, where the maps $P\colon \mathscr{V}^{(3)}(G) \to \mathscr{V}(G)$ and $\Lambda : \mathscr{V}^{(3)}(G) \to \mathscr{V}(G)$ are induced on global sections by the sheaf morphisms p and λ.*

More simply, the graph Laplacian can be implemented as a topological filter.

Proof The result follows from the observation that the space of global sections $\mathscr{V}^{(3)}(G)$ is isomorphic to $\mathbb{R}^N$, where N is the number of vertices in the graph, and that the i-th column of L can be read directly off the construction of λ_{v_i}. □

3.4.3 Thresholding Filters

In addition to linear filters, topological filters are a good context for studying certain *nonlinear* filters. Although vitally important in applications, thresholding filters have a rather different mathematical basis than linear filters, and are usually treated specially. Since both are topological filters, sheaves provide a context for treating both kinds of filters uniformly.

Example 3.25 (compare with Example 4.6) The canonical example of a nonlinear filter is the thresholding filter. We can encode this as a topological filter from the sheaf of continuous functions to a sheaf of open sets. It suffices to consider a single morphism of sheaves of $\mathbb{R}$, as given by the diagram

$$\begin{array}{ccccccc} \cdots \longrightarrow & C((-1,0)) & \longleftarrow & C((-1,1)) & \longrightarrow & C((0,1)) & \longleftarrow \cdots \\ & \downarrow & & \downarrow & & \downarrow & \\ \cdots \longrightarrow & \mathscr{T} \cap (-1,0) & \xleftarrow[U \mapsto U \cap (-1,0)]{} & \mathscr{T} \cap (-1,1) & \xrightarrow[U \mapsto U \cap (0,1)]{} & \mathscr{T} \cap (0,1) & \longleftarrow \cdots \end{array}$$

in which $\mathscr{T}$ is the collection of open sets of $\mathbb{R}$, and each downward arrow is given by $f \mapsto \{x : f(x) > T\}$. Observe that while $\mathscr{T}$ is usually given no topological structure, $C((a,b))$ is usually given a topology (the compact-open topology, for instance). (Indeed, the downward maps $C((n, n+1)) \to \mathscr{T} \cap (n, n+1)$ are not continuous in any of the commonly used topologies on sets of subsets.)

As an example of how this detector performs, consider the diagram

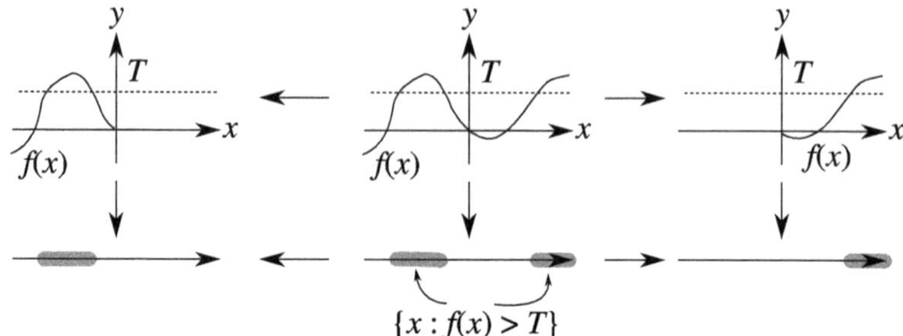

which shows the progression of a function through the maps in the filter. The maps to the left restrict the consideration from the function on $(-1, 1)$ to $(-1, 0)$. The detected region is represented by the set of points where the function exceeds the threshold T. This set is restricted as well.

Exercise 3.9 Extend the diagrams in Example 3.25 to the left and right. Hint: on the immediate left of the existing diagrams, consider the values of the function on $(-2, -1)$.

In practice, the thresholding filter in Example 3.25 is too simplistic. The threshold should be varied to account for a varying noise floor. If the threshold is chosen to be proportional to a local average of the function, then the resulting detector locates where the signal is anomalously large. This is called a *constant false alarm rate* (CFAR) filter. Such a filter can be constructed from a topological filter that is based on the thresholding filter constructed in the previous example.

As before, let $\mathscr{V}$ be the $\mathbb{R}$-sampling sheaf supported on vertices $\mathbb{Z} \subset \mathbb{R}$ for the usual cell complex structure on $\mathbb{R}$. Let $\mathscr{V}^{(M)}$ be the M-term grouping sheaf with the morphism $p : \mathscr{V}^{(M)} \to \mathscr{V}$ whose component maps are given by the projection in Eq. 3.3, again as before. The output of the CFAR filter will be placed in a sheaf $\mathscr{B}$ that is a binary sampling sheaf, given by the diagram

$$\cdots \longrightarrow 0 \longleftarrow \{0, 1\} \longrightarrow 0 \longleftarrow \{0, 1\} \longrightarrow \cdots$$

The CFAR filter is then implemented by the following topological filter

$$\mathscr{V} \xleftarrow{\;p\;} \mathscr{V}^{(M)} \xrightarrow{\;\tau\;} \mathscr{B},$$

where τ is defined by its component maps $\tau_v : \mathbb{R}^M \to \{0, 1\}$ on a vertex v. Each of these component maps are given by

$$\tau_v(a_1, \dots, a_M) = \begin{cases} 0 \text{ if } a_{[M/2]} \le (1/M) \sum_{i=1}^{M} a_i + T_{\text{offset}} \\ 1 \text{ if } a_{[M/2]} > (1/M) \sum_{i=1}^{M} a_i + T_{\text{offset}} \end{cases}$$

where T_{offset} is a parameter that adjusts the sensitivity of the filter. Larger values of T_{offset} make the filter less sensitive, while smaller values make it more

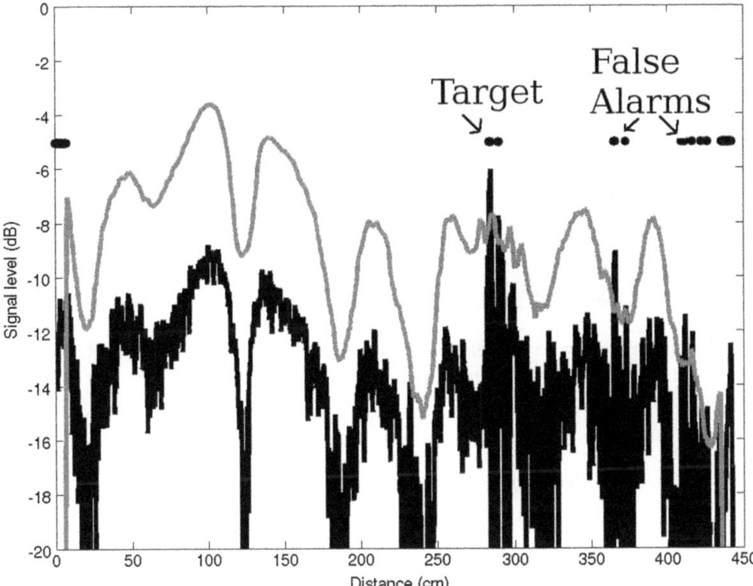

Fig. 3.14 Thresholding a sonar signal (*black*) using the detection threshold (*gray*) to result in detections (*black dots*)

sensitive. In implementations, the quantity $(1/M)\sum_{i=1}^{M} a_i + T_{\text{offset}}$ is called the *detection threshold*.

As an example of the CFAR process, Fig. 3.14 shows sonar echos collected by the author in which a reflective target was located at a range of 280 cm from the sensor. The input signal is shown in black, and represents acoustic signal strength as a function of distance to the sensor. This input signal is represented in the sheaf model as a global section of $\mathscr{V}$. The signal was collected using an acoustic horn antenna and digitized by a laptop sound card at a sampling rate of 44.1 kHz. Blocks of 100 adjacent samples (corresponding to about about 2.26 ms) were averaged to form the detection threshold, which includes an offset of 6 dB. The resulting threshold is shown in gray on Fig. 3.14. Distances where the signal strength exceeded this threshold are shown as large dots at the -5 dB level, and correspond to the places where the global section of $\mathscr{B}$ takes the value 1. The target's echo is clearly visible, but the signal exceeded the threshold at other locations as well. These other locations are called *false alarms* and are probably due to other reflective obstacles in the scene.

Remark 3.6 The component maps τ_v can be "tuned" considerably to reduce false alarms. For instance, many implementations construct a detection threshold of the form

Fig. 3.15 An image (482 × 653 pixels) with curved striations (*left*) and its LSRA filtered image (*right*), in which the colors represent angle in degrees. The filter used a block size of 30 pixels and a spectral radius between 3 and 8 pixels

$$\frac{1}{[M/2]-N}\sum_{i=1}^{[M/2]-N} a_i + \frac{1}{M-N-[M/2]}\sum_{i=[M/2]+N}^{M} a_i + T_{\text{offset}},$$

which avoids biasing the threshold with potentially anomalous values. Other implementations weight the terms in the above sums. The interested reader is encouraged to consult Stimson (1998) for details about practical CFAR implementations.

3.4.4 Angle-Valued Filters

Curved striations are sometimes an important feature to be detected in an image. For instance, the left panel of Fig. 3.15 shows a photograph of a stack of dishes. The collection of edges of the dishes forms a striated feature in the image. It is therefore useful to have a filter that measures the orientation of striated features from an image. It is most effective to describe this orientation by an angle.

In this section, we describe a topological filter called the local spectral rotation angle (LSRA) which takes an intensity-valued image to an angle-valued image. Of necessity, this filter will not be linear, since the space of angles is not a vector space (it is a group). On the other hand, it is local since the orientation of striations should be allowed to change across the image.

Topological filters provide a solid foundation on which to construct a method for making local angular measurements of striations in an image. Essentially, the desired filter should compute the angle of any striations in small patches of an image, and then assemble the resulting computations into an angle-valued image, as shown in Figure 3.16. The topological filter described here uses (1) the 2d-Fast Fourier Transform of a small patch followed by (2) a threshold detection on an annular window to determine the dominant angle.

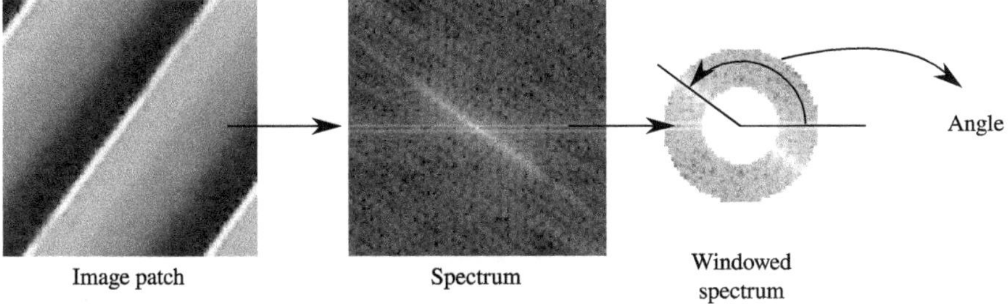

Fig. 3.16 Schematic of the local spectral angle calculation

The LSRA topological filter takes as input a rectangular grid of $\mathbb{R}$-valued pixels (the input image) and four parameters:

1. M, the number of rows in a local patch,
2. N, the number of columns in a local patch,
3. R_1, the minimum spectral radius to test, and
4. R_2, the maximum spectral radius to test.

It produces a new rectangular grid of pixels, for which the value of each pixel is an angle.

Specifically, for a $M \times N$ patch s of pixels, the local process computes its discrete Fourier transform

$$S(k_x, k_y) = \frac{1}{MN} \sum_{x=0}^{N-1} \sum_{y=0}^{M-1} e^{2\pi i(k_x x/N - k_y y/M)} s_{x,y}$$

for $k_x = -N/2, \ldots, N/2$, $k_y = -M/2, \ldots, M/2$. The local process then computes the frequency coordinates (T_x, T_y) of the largest spectral component in an annulus via

$$(T_x, T_y) = \underset{R_1 \le \sqrt{k_x^2 + k_y^2} \le R_2}{\operatorname{argmax}} |S(k_x, k_y)|$$

and returns the angle $\theta_{x,y} = \tan^{-1} \frac{T_y}{T_x}$ that the dominant spectral component makes with the horizontal. By convention, we assume that $-\pi/2 \le \theta_{x,y} < \pi/2$. Note that R_1 and R_2 are the spectral radius parameters that are specified in advance.

It is important to realize that the direction of the striations themselves is perpendicular to the dominant spectral components. As Fig. 3.16 indicates, the striations (left panel) make an angle of roughly 45° with the horizontal, but the spectral band makes an angle of −45° with the horizontal. In this way, the local process implements a function $A : \mathbb{R}^{MN} \to S^1$ taking the image patch to S^1, the unit circle.

The construction of the LSRA filter as a topological filter proceeds much as in the previous sections. We encode the input image as a sheaf $\mathscr{V}$ with diagram

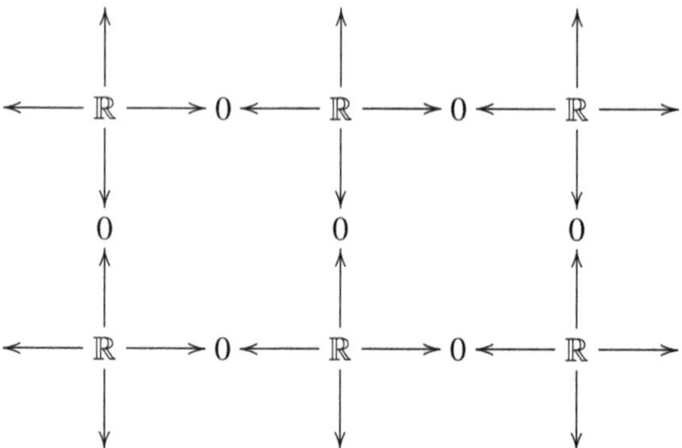

which is an $\mathbb{R}$-sampling sheaf on a square grid. Similarly, we define $\mathscr{S}$ to be the S^1-sampling sheaf on the same square grid. We construct a grouping sheaf $\mathscr{V}^{(M,N)}$ in much the same way as for a linear shift-invariant system, so that it has diagram

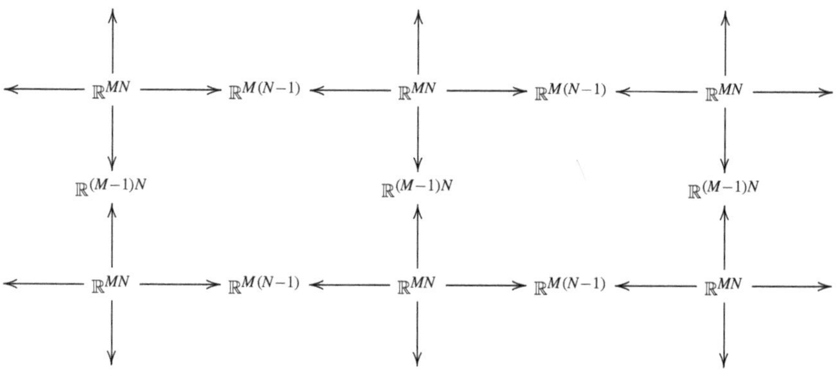

The LSRA is then the topological filter

$$\mathscr{V} \xleftarrow{p} \mathscr{V}^{(M,N)} \xrightarrow{\alpha} \mathscr{S},$$

where p has component maps that are projections $\mathbb{R}^{MN} \to \mathbb{R}$ onto the first component for each vertex, and α has component maps that are the function A on each vertex.

Exercise 3.10 Complete the construction of $\mathscr{V}^{(M,N)}$ by specifying the restrictions and show that $p \colon \mathscr{V}^{(M,N)} \to \mathscr{V}$ induces an isomorphism on spaces of global sections.

Figures 3.15 and 3.17 show a collection of three grayscale photographs (left panels), each of which are 482×653 pixels. Each such image contains striations at different angles, and Fig. 3.15 contains curved striations. The LSRA was applied to

Fig. 3.17 Two example images (each 482 × 653 pixels) (*left*) and their local spectral angle transforms (*right*) using a block size of 30 × 30 pixels and a spectral radius of between 3 and 8 pixels

each photograph (right panels) using a block size $M = N = 30$ pixels, and a spectral radius between $R_1 = 3$ and $R_2 = 8$ pixels.

The specific angles produced by the LSRA correspond to the orientation of the visually apparent striations, accounting for the addition of $\pm 90°$ as described earlier. For instance, consider the lower left panel of Fig. 3.17, in which the striations (the window blinds) are horizontal. This corresponds to an angle of $0°$ made with the horizontal, which yields a spectral angle of $\pm 90°$. This spectral angle is what is visible in the the lower right frame, though there is some oscillation between $+90°$ (white) and $-90°$ (black). If the image is rotated, as in the upper frames of Fig. 3.17, the corresponding angle measured by the LSRA changes by the applied rotation angle as well.

In Fig. 3.15, the angles of the striations along the edge of the dishes progress from roughly $+45°$ (left side) to $-45°$ (right side), passing through the branch cut at $\pm 90°$ along the way. Regions in which the striations are oriented in the same direction are colored similarly, though there is some amount of "quantization." This results in the discontinuities in the center of the lower right panel of Figs. 3.17 and 3.15.

These examples indicate that the LSRA is an effective filter for measuring angles in an image. Its robustness can be attributed to the fact that it is *local*; an error in angle estimation disrupts only a limited portion of the image. The locality of the filter also means that it retains much of the spatial resolution of the input image, and that the resolution loss that does occur can be controlled by the choice of block size.

Fig. 3.18 Representation of a dense urban environment (*left*, building geometry from Mannesmann Mobilfunk GmbH (1999)) as a graph with edges weighted by length (*right*)

We close this case study by noting that the LSRA is one of many related topological filters that perform local angle computations.

3.5 Case Study: Indoor Wave Propagation

One of the first places where signal processing got its start was in the transmission of signals along telegraph wires. Traveling waves along strings or wires usually involve the wave equation

$$\frac{\partial^2 u}{\partial t^2}(t,x) = c^2 \frac{\partial^2 u}{\partial x^2}(t,x) \tag{3.4}$$

with wave speed c, or the Helmholtz equation

$$\frac{\partial^2 U}{\partial x^2}(x) + k^2 U(x) = 0,$$

which arises if $u(t,x) = U(x)e^{ikct}$, where k is the *wavenumber*. This remains a good approximation if wave propagation happens along narrow channels, though the domain is better represented as a graph G instead of $\mathbb{R}$. (See Kostrykin (1999); Kuchment (2002); Smilansky (2006); Molchanov (2006) for precise conditions under which this approximation holds.) This situation often arises in the study of radio propagation in urban canyons, for instance in the portion of Munich shown in Fig. 3.18.

Since the lengths of edges in such a graph G play an important role in the propagation of waves, an important inverse problem is the reconstruction of these lengths from measurements of a small number of waves. From an urban propagation perspective, the reconstruction of edge lengths from signal measurements is a form of

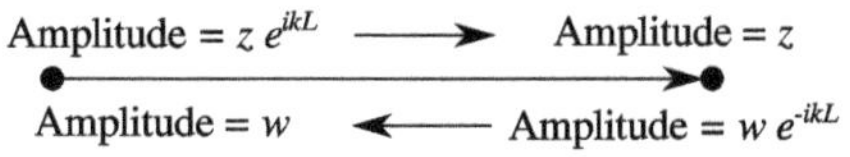

Amplitude = $z\,e^{ikL}$ ⟶ Amplitude = z

Amplitude = w ⟵ Amplitude = $w\,e^{-ikL}$

Fig. 3.19 A section of a transmission line sheaf supported with value (w, z) on an edge with length L

1-dimensional topological "imaging" that produces a geometric map of an environment. As in Sect. 2.3, this map is somewhat more abstract and qualitative than what traditional imaging algorithms produce.

Existing approaches for performing this inversion usually rely on wideband spectral methods, which examine eigenfunctions of a differential operator on G that generalizes the Helmholtz operator $\left(\frac{\partial^2}{\partial x^2} + k^2\right)$. This has unfortunate consequences for most urban sensing applications, since the available bandwidth for sensing is limited.

In the context of wave propagation on a graph, the space of solutions forms a sheaf. This even holds if the propagation is lossy, or if we instead consider fundamental solutions, in which there are a number of sources (Robinson 2010). In this case study, we show that sheaves provide a framework for a narrowband sensing methodology.

We will assume that the particular cell complex structure of the graph is known, and then use measurements of waves on the graph to reconstruct the edge lengths. This is not as limited as it might seem, since an algorithm is presented in Sect. 4.7 for reconstructing this assumed topological structure.

3.5.1 Transmission Line Sheaves

Suppose that X is a 1-dimensional cell complex representing the propagation environment. Recall that the one-point compactification of X is $\underline{X} = X \sqcup \{*\}$, which is a CW complex. Let us suppose that each edge e of $\underline{X}$ whose closure does not intersect $*$ is assigned a positive real number $L(e)$, called its *length*. Edges whose closure does intersect $*$ will be called *external edges* and will be assigned a length of zero.

Definition 3.11 Suppose X is a 1-dimensional cell complex X whose edges e are labeled by (1) a length L as above and (2) an arbitrary direction. The *transmission line sheaf* $\mathscr{T}$ with wavenumber k is given by

1. $\mathscr{T}(v) = \mathbb{C}^{\deg v}$ for all vertices v,
2. $\mathscr{T}(e) = \mathbb{C}^2$ for all edges e, and
3. If e_m is the m-th edge attached to a degree n vertex v,

$$\mathscr{T}(v \rightsquigarrow e_m)(u_1, \dots, u_n) = \begin{cases} \left(u_m, e^{-ikL(e_m)}\left(\frac{2}{n}\sum_{j=1}^n u_j - u_m\right)\right) & \text{if } e_m \text{ is inward at } v \\ \left(e^{ikL(e_m)}\left(\frac{2}{n}\sum_{j=1}^n u_j - u_m\right), u_m\right) & \text{if } e_m \text{ is outward at } v \end{cases}$$

The stalks of a transmission line sheaf represent the complex amplitudes of the travelling waves *exiting* each of its two endpoints, measured at the endpoints. Because each edge has an assumed orientation, the first component represents the wave amplitude at the outgoing vertex, while the second component represents the wave amplitude at the incoming vertex. It is important to note that the complex amplitudes measured at other points will be phase shifted with respect to these points. On an edge with length L, a travelling wave that exits with amplitude z will have complex amplitude ze^{ikL} at the entrance. Figure 3.19 summarizes this interpretation of the stalks over an edge in a transmission line sheaf. In contrast, the stalk over a vertex represents the complex wave amplitudes of waves *entering* the vertex.

Example 3.26 The real line can be realized as a single edge, which results in a transmission line sheaf whose global sections are parametrized by $\mathbb{C}^2$ even though this is not a cell complex structure for $\mathbb{R}$. They correspond to left-moving and right-moving waves, $u_1 e^{ikx} + u_2 e^{-ikx}$, respectively.

The real line can also be realized as a cell complex with one vertex and two external edges. In this case, the transmission line sheaf has diagram

$$\mathbb{C}^2 \xleftarrow{f^-} \mathbb{C}^2 \xrightarrow{f^+} \mathbb{C}^2$$

where

$$f^+(u_1, u_2) = (u_1, u_2) \text{ and } f^-(u_1, u_2) = (u_2, u_1).$$

Clearly, the space of global sections is still $\mathbb{C}^2$.

The global sections of a transmission line sheaf correspond to solutions to the Helmholtz equation on each edge of X that satisfy the following *Kirchhoff conditions* at each vertex v

1. All solutions on the edges attached to v extend to a single continuous function at v
2. The sum of the derivatives (taken in the direction away from v) of each solution at v is zero.

Exercise 3.11 Show that the two conditions above lead to the definition of the restriction map for a transmission line sheaf.

Example 3.27 Consider the floorplan of the third floor of the David Rittenhouse Laboratory of the University of Pennsylvania shown in Fig. 3.20. We can realize the structure relevant to propagation in a cell complex, as shown on the left of Fig. 3.21. In that figure, arbitrary orientations have been assigned to each edge.

Using Definition 3.11, a transmission line sheaf can be constructed on this cell complex model. The diagram of this sheaf is shown at right of Fig. 3.21.

The restrictions are given by the following matrices:

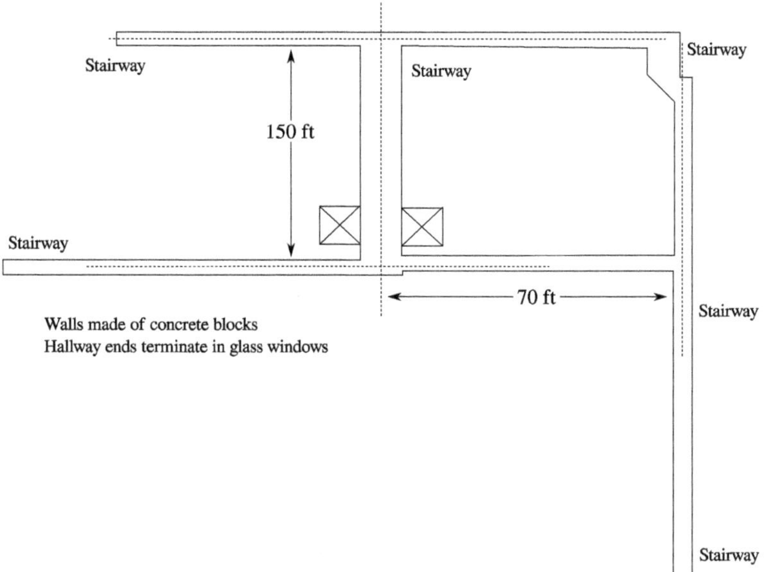

Fig. 3.20 Dimensioned floorplan of the third floor of David Rittenhouse Laboratory

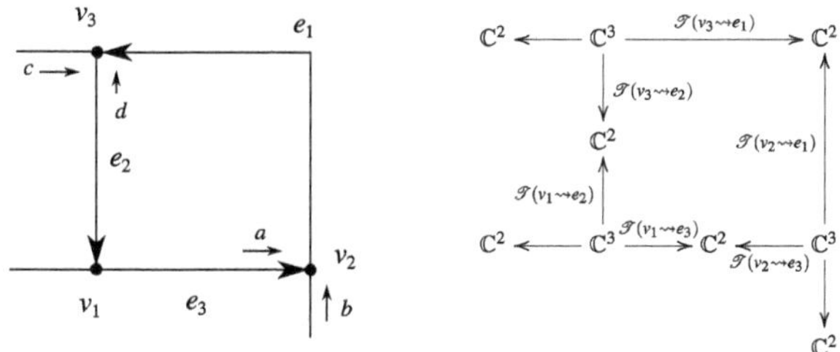

Fig. 3.21 Directed graph (*left*) and transmission line sheaf (*right*) for the third floor of David Rittenhouse Laboratory

$$\mathscr{T}(v_1 \rightsquigarrow e_2) = \begin{pmatrix} -\frac{1}{3}e^{ikL(e_2)} & \frac{2}{3}e^{ikL(e_2)} & \frac{2}{3}e^{ikL(e_2)} \\ 1 & 0 & 0 \end{pmatrix}$$

$$\mathscr{T}(v_1 \rightsquigarrow e_3) = \begin{pmatrix} 0 & 1 & 0 \\ \frac{2}{3}e^{-ikL(e_3)} & -\frac{1}{3}e^{-ikL(e_3)} & \frac{2}{3}e^{-ikL(e_3)} \end{pmatrix}$$

$$\mathscr{T}(v_2 \rightsquigarrow e_1) = \begin{pmatrix} 0 & 1 & 0 \\ \frac{2}{3}e^{-ikL(e_1)} & -\frac{1}{3}e^{-ikL(e_1)} & \frac{2}{3}e^{-ikL(e_1)} \end{pmatrix}$$

$$\mathscr{T}(v_2 \rightsquigarrow e_3) = \begin{pmatrix} -\frac{1}{3}e^{ikL(e_3)} & \frac{2}{3}e^{ikL(e_3)} & \frac{2}{3}e^{ikL(e_3)} \\ 1 & 0 & 0 \end{pmatrix}$$

$$\mathscr{T}(v_3 \rightsquigarrow e_1) = \begin{pmatrix} -\frac{1}{3}e^{ikL(e_1)} & \frac{2}{3}e^{ikL(e_1)} & \frac{2}{3}e^{ikL(e_1)} \\ 1 & 0 & 0 \end{pmatrix}$$

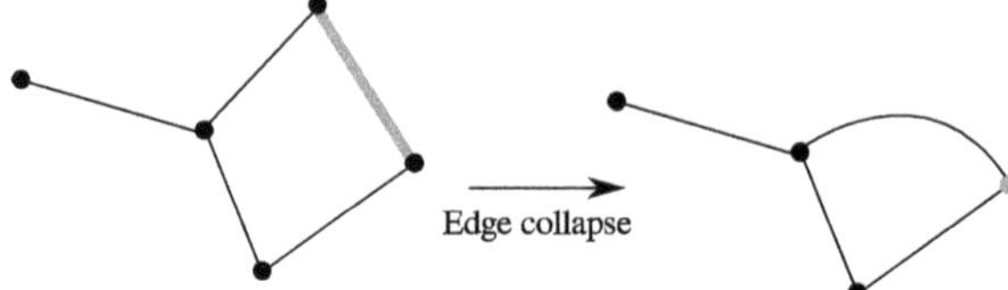

Fig. 3.22 The collapse of an edge (*marked in gray*) of a cell complex

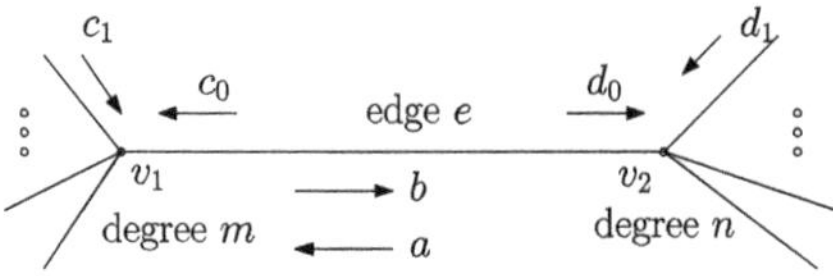

Fig. 3.23 Vertices v_1 and v_2 incident to an edge e to be collapsed

$$\mathscr{T}(v_3 \rightsquigarrow e_2) = \begin{pmatrix} 0 & 1 & 0 \\ \frac{2}{3}e^{-ikL(e_2)} & -\frac{1}{3}e^{-ikL(e_2)} & \frac{2}{3}e^{-ikL(e_2)} \end{pmatrix}$$

Exercise 3.12 Following Definition 3.11, construct the restriction maps from the vertices to the external edges in Example 3.27.

3.5.2 Sheaf Pushforwards and Edge Collapse

Transmission line sheaves behave particularly well under certain pushforwards. This allows their structure to be simplified dramatically. The precise condition under which the pushforward operation preserves a transmission line sheaf occurs when a cellular map *collapses* an edge. Specifically, a cellular map $f : X \to Y$ is an *edge collapse* (see Fig. 3.22) if it is a homeomorphism on all of X except a single edge e with distinct endpoints which f takes to a single vertex.

Proposition 3.6 *Suppose that $\mathscr{T}$ is a transmission line sheaf on X and that $f : X \to Y$ is an edge collapse. Then f_* is an isomorphism of sheaves.*

Proof Suppose that the endpoints of e are v_1 and v_2, with total degrees m and n respectively. Let the stalk over v_1 be spanned by $c_0, \ldots, c_{m-1}$, the stalk over v_2 be spanned by $d_0, \ldots, d_{n-1}$, and the stalk over e be spanned by a, b, as shown in Fig. 3.23.

It suffices to show (1) that the space of sections over v_1, v_2, and e is the same as the space of sections over a vertex with degree $m + n - 2$ and (2) that restriction maps of $f_*\mathscr{T}$ at $f(e)$ satisfy the axioms for a transmission line sheaf up to an isomorphism. (In particular this means that the pushforward of the sheaf over v_1, v_2, e is an isomorphism.)

1. We compute the space of sections directly, by equating the restrictions of $\mathcal{T}$ from v_1 and v_2 to e. Namely,

$$a = e^{-ikL(e)} \left(\frac{2}{n} \sum_{j=1}^{n-1} d_j + \left(\frac{2}{n} - 1 \right) d_0 \right) = c_0 \tag{3.5}$$

and

$$b = e^{ikL(e)} \left(\frac{2}{m} \sum_{j=1}^{m-1} c_j + \left(\frac{2}{m} - 1 \right) c_0 \right) = d_0,$$

where $L(e)$ is the length of edge e. Substituting the equation for d_0 yields an equation for c_0 in terms of $\{c_1, \ldots, c_{m-1}\}$ and $\{d_1, \ldots, d_{n-1}\}$. Hence, the dimension of the space of global sections is parametrized by these two sets, and therefore is $m + n - 2$. This is the same dimension as the stalk over $f(e)$ should have in a transmission line sheaf.
2. Continuing with the substitution in (1), we obtain

$$c_0 = \frac{me^{-ikL(e)}}{m+n-2} \sum_{j=1}^{n-1} d_j + \frac{2-n}{m+n-2} \sum_{j=1}^{m-1} c_j. \tag{3.6}$$

Now let us verify that the pushforward sheaf is indeed a transmission line sheaf. Suppose without loss of generality that the p-th edge connected to $f(e)$ was originally connected to v_1. Restricting to this edge yields one trivial component, and one which has the form

$$e^{ikL(e_p)} \left(\frac{2}{m} \sum_{j=0}^{m-1} c_j - c_p \right).$$

Substituting (3.6) for c_0, this reduces to

$$e^{ikL(e_p)} \left(\frac{2e^{ikL(e)}}{m+n-2} \sum_{j=1}^{n-1} d_j + \frac{2}{m+n-2} \sum_{j=1}^{m-1} c_j - c_p \right),$$

which is almost exactly the form given in Definition 3.11 for a degree $m+n-2$ vertex, except for the presence of the factor of $e^{ikL(e)}$. □

Example 3.28 Continuing Example 3.27, consider the cellular map f which collapses edge e_1. Let $\mathcal{S} = f_* \mathcal{T}$, whose diagram is shown in Fig. 3.24.

Using the calculations in the proof of Propostion 3.6, we have that the stalk $\mathcal{S}(v_2)$ has dimension 4, for which we assume a basis $\{a, b, c, d\}$ (in that order) as shown

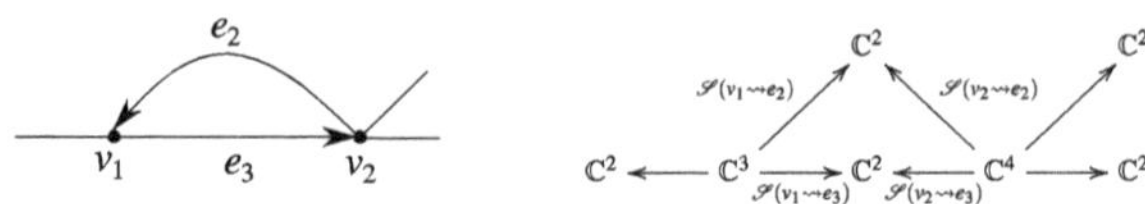

Fig. 3.24 Directed graph (*left*) and transmission line sheaf (*right*) in Fig. 3.21 after collapsing edge e_1

in Fig. 3.21. The restrictions in the collapsed graph then are given by the following (notice the presence of the $L(e_1)$ terms)

$$\begin{aligned}
\mathscr{S}(v_1 \rightsquigarrow e_2) &= \begin{pmatrix} -\frac{1}{3}e^{ikL(e_2)} & \frac{2}{3}e^{ikL(e_2)} & \frac{2}{3}e^{ikL(e_2)} \\ 1 & 0 & 0 \end{pmatrix} \\
\mathscr{S}(v_1 \rightsquigarrow e_3) &= \begin{pmatrix} 0 & 1 & 0 \\ \frac{2}{3}e^{-ikL(e_3)} & -\frac{1}{3}e^{-ikL(e_3)} & \frac{2}{3}e^{-ikL(e_3)} \end{pmatrix} \\
\mathscr{S}(v_2 \rightsquigarrow e_2) &= \begin{pmatrix} 0 & 0 & 0 & 1 \\ \frac{1}{2}e^{ik(L(e_1)-L(e_2)} & \frac{1}{2}e^{ik(L(e_1)-L(e_2))} & \frac{1}{2}e^{-ikL(e_2)} & -\frac{1}{2}e^{-ikL(e_2)} \end{pmatrix} \\
\mathscr{S}(v_2 \rightsquigarrow e_3) &= \begin{pmatrix} -\frac{1}{2}e^{ikL(e_3)} & \frac{1}{2}e^{ikL(e_3)} & \frac{1}{2}e^{ik(L(e_3)-L(e_1))} & \frac{1}{2}e^{ik(L(e_3)-L(e_3))} \\ 1 & 0 & 0 & 0 \end{pmatrix}.
\end{aligned}$$

Proposition 3.6 is false when e does not have distinct endpoints. For instance, a transmission line sheaf on a graph with a loop will have different global sections than one on a simply connected graph.

Example 3.29 Consider the CW complex structure on the circle with one vertex v and one edge e of length L. In this case, the global sections of a transmission line sheaf are characterized by the equations

$$u_1 = e^{ikL}u_1 \text{ and } u_2 = e^{-ikL}u_2.$$

This means that the space of global solutions is either the trivial vector space if $kL \notin 2\pi\mathbb{Z}$ or is $\mathbb{C}^2$ if $kL \in 2\pi\mathbb{Z}$. In contrast, the space of global sections of a transmission line sheaf on a single vertex is trivial. The case where the loop has more global sections (when $kL \in 2\pi\mathbb{Z}$) is called *resonance*.

A global section of a sheaf is uniquely specified by its value on all vertices. In the case of a transmission line sheaf, a global section constrains the geometry of the graph. The calculation of the values of a section along the collapsed edge in Proposition 3.6 is a useful tool for "sounding" the length of edges in a graph. One needs to place a directional sensor at each vertex in the graph and measure the incoming wave amplitudes. For instance, a horn antenna could be placed at each intersection of hallways in a building, and oriented in the direction of each hallway for each measurement. Then, the following Algorithm can be recursively called, once for each edge with an unknown length.

Algorithm 1 (Extracting the geometry of a graph by sounding)
Input:

1. Graph model in which the edges are all marked with lengths except edge e, whose length is unknown.
2. Complex signal measurements $c_0, d_0, \dots, d_{n-1}$ taken from the ends of edge e, and collected at several algebraically independent operating wavenumbers.

Output: $L(e)$, the length of edge e.
Procedure:

1. Evaluate
$$L(e) = \frac{i}{k} \log \left(\frac{1}{c_0} \left(\frac{2}{n} \sum_{j=1}^{n-1} d_j + \left(\frac{2}{n} - 1 \right) d_0 \right) \right). \tag{3.7}$$
for each wavenumber available, which is merely Eq. (3.5) solved for $L(e)$.
2. Select the correct branch of the logarithm as being the one yielding a length consistent with each of the available wavenumbers.

Example 3.30 Consider the case of attempting to measure the geometry of Fig. 3.20 from the graph model in Fig. 3.21. Suppose that three sensors are placed at each of v_1, v_2, and v_3, for which simulated magnitude and phase measurements shown in Table 3.1. These measurements correspond to two operating frequencies, one at 905 MHz and one at 2.451 GHz (typical wireless network frequencies), and were simulated by solving the lossless Helmholtz equation on a graph in which the edge lengths were as shown in Fig. 3.20, namely $L(e_1) = 220$ ft, $L(e_2) = 150$ ft, and $L(e_3) = 70$ ft. The choice of frequences is important; frequencies with small common factors make accurate measurements of edge lengths difficult.

Using the principal branches of Eq. 3.7 allows us to estimate the lengths of e_1, e_2, and e_3. For instance, using the operating frequency of 905 MHz yields the following estimates:

- $L(e_1) \approx -0.233$ ft, which is too small by exactly 405 half-wavelengths,
- $L(e_2) \approx 0.459$ ft, which is too small by exactly 275 half-wavelengths, and
- $L(e_3) \approx -0.148$ ft, which is too small by exactly 129 half-wavelength.

Using 2.451 GHz alone isn't more accurate, since it yields the following estimates:

- $L(e_1) \approx 0.140$ ft, which is too small by exactly 1095 half-wavelengths,
- $L(e_2) \approx 0.0130$ ft, which is too small by exactly 747 half-wavelengths, and
- $L(e_3) \approx -0.0742$ ft, which is too small by exactly 349 half-wavelengths.

However, combining the two frequencies alleviates the difficulty, in the following way. Suppose we have two estimates L and L' for an edge length, associated to wavelengths λ and λ', we then search for the smallest such value that satisfies

Table 3.1 Simulated magnitude and phase measurements for Example 3.30

Vertex	Hallway	Mag at 905 MHz (dB)	Phase at 905 MHz (°)	Mag at 2451 MHz (dB)	Phase at 2451 MHz (°)
v_1	External	1.4	−8.6	0.44	−14
v_1	e_2	−5.5	−1.5	−0.82	−28
v_1	e_3	−6.6	82	−4.9	83
v_2	External	1.9	−78	0.57	−85
v_2	e_1	−3.4	156	−1.7	108
v_2	e_3	−6.6	131	−4.9	150
v_3	External	0	0	0	0
v_3	e_1	−3.4	−126	−1.7	−17
v_3	e_2	−5.5	150	−0.82	−16

$$L + m\frac{\lambda}{2} = L' + n\frac{\lambda'}{2},$$

where m and n are integers. Performing this search on each of the edges in our graph yields the correct lengths, namely $L(e_1) \approx 220$ ft, $L(e_2) \approx 150$ ft, and $L(e_3) \approx 70$ ft.

3.6 Open Questions

1. Angle-valued image filters have only recently begun to be studied. The desirable or undesirable performance characteristics have not yet been carefully described. What constitutes signal-to-noise ratio for angle-valued filters, since the space of angles (the unit circle) does not have a linear order? It is likely that some partial order may suffice, but it is yet unclear how to make that assessment.
2. The topological filters described in this chapter have all made use of grouping sheaves as their internal state. This is clearly not necessary to the construction; what other useful classes of topological filters are there?
3. Signal propagation on a cell complex generalizes the notion of a quantum graph to higher dimensional spaces. It is not difficult to write a description of a sheaf of solutions, given reasonable matching conditions along lower-dimensional cells. However, for complexes with cells of dimension 2 and higher, this sheaf has infinite-dimensional stalks described by function spaces. What filtration of finite-dimensional subsheaves can one use to approximate these sheaves and their spaces of global sections? We want the representation to be well-behaved with respect to the asymptotic expansions that give rise to geometric optics and diffraction!
4. Networks are an area of recent interest; communication and social networks are often only loosely associated to physical locations, and so frustrate traditional analyses. However, the agents involved in these networks do maintain some notion of locality, for instance a collection of "friends." Because of this, many researchers believe that graphs (or 1d cell complexes) are appropriate for modeling the

topological structure of a network. However, the true importance of a network lies in the information it carries; therefore it seems that sheaves might be ideal for studying networks. Several researchers Lilius (1993); Ghrist (2011); Robinson (2012) have found specific networks that admit a straightforward sheaf model. Is there a general sheaf-theoretic model of networks?

References

Bacławski K (1975) Whitney numbers of geometric lattices. Adv Math 16:125–138

Bredon G (1997) Sheaf theory. Springer, Berlin

Ghrist R, Hiraoka Y (2011) Applications of sheaf cohomology and exact sequences to network coding. preprint

Godement R (1958) Topologie algebrique et théorie des faisceaux. Herman, Paris

Hubbard JH (2006) Teichmüller theory, vol 1. Matrix Editions, Ithaca, NY

Iverson B (1984) Cohomology of sheaves. Aarhus universitet, Matematisk institut, Aarhus

Kostrykin V, Schrader R (1999) Kirchoff's rule for quantum wires. J Phys A Math Gen 32:595–630

Kuchment P (2002) Graph models for waves in thin structures. Waves Random Complex Media 12(4):R1–R24

Lilius J (1993) Sheaf semantics for Petri nets. Technical report, Helsinki University of Technology, Digital Systems Laboratory, Espoo

Mannesmann Mobilfunk GmbH (1999) Cost231 urban micro cell measurements and building data, originally http://www2.ihe.uni-karlsruhe.de/forschung/cost231/cost231.en.html, now defunct

Molchanov S, Vainberg B (2006) Transition from a network of thin fibers to the quantum graph: an explicitly solvable model. In: Quantum graphs and their applications. Contemp Math 18(415):227–240

Robinson M (2010) Inverse problems in geometric graphs using internal measurements, arxiv:1008.2933

Robinson M (2012) Asynchronous logic circuits and sheaf obstructions. Electron Notes Theor Comput Sci 283:159–177

Shepard A (1980) A cellular description of the derived category of a stratified space, Ph.D. thesis. Brown University, Providence

Smilansky U, Solomyak M (2006) The quantum graph as a limit of a network of physical wires. In: Quantum graphs and their applications, pp 283–292

Stimson GW (1998) Introduction to Airborne radar. Scitech Publishing, Inc., Mendham, NJ

Chapter 4
Detection

This chapter will explain

1. That typical detector algorithms are instances of *functors*,
2. The difference between detectors and filters by their effect on signals,
3. The class of detectors called *sheaf cohomology*, which compute global consistency relationships between measurements,
4. That sampling theory is generalized by the use of *exact sequences* for cohomology, and

will exhibit a direct, algorithmic application of cohomology that permits consistent assembly of multi-sensor measurements of visibility.

Detectors are tools which extract and emphasize important features of a signal. In order to be useful, a detector should be *functorial*, in that it preserves the features of the signal. Which features are important may change depending on the context, so the appropriate detector is also context-dependent.

In most practical settings, it is desirable for detectors to lose information. Specifically, one usually wants to transform a signal into a representation that removes noise and irrelevant signals. Merely requiring that certain qualitative features of a signal be preserved will motivate our definition of detectors as certain kinds of *forgetful functors*. The most famous detector for sheaves of abelian groups is *sheaf cohomology*. We will examine applications of cohomology to sampling theory and urban radio mapping.

4.1 Categories and Functors

Signal processing posits that the world contains objects which are related to one another, and that measurements can be used to make inferences about these objects and their relationships. If this is to be a fruitful endeavor, the relationships between measurements should be similar to the relationships between the objects they measure. This idea is distilled into the mathematical notion of a *category*.

M. Robinson, *Topological Signal Processing*,
Mathematical Engineering, DOI: 10.1007/978-3-642-36104-3_4,

Definition 4.1 A *category* **C** is a class of *objects* **Obj**(**C**) and a class of *morphisms* **Mor**(**C**) that satisfy the following axioms:

1. Each morphism $m \in \mathbf{Mor}$ has a unique *source* and *target*, both of which are in **Obj**. If the source of m is $a \in \mathbf{Obj}$ and its target is $b \in \mathbf{Obj}$, we write $m : a \to b$.
2. (Morphisms are composable) There is a binary operation on morphisms called *composition* that for every pair of morphisms $m : a \to b$ and $n : b \to c$ (with $a, b, c \in \mathbf{Obj}$) there is a unique morphism $n \circ m : a \to c$. In this case, m and n are said to be *composable*.
3. (Composition is associative) If $m : a \to b$, $n : b \to c$, and $p : c \to d$ are three morphisms, then $(p \circ n) \circ m = p \circ (n \circ m)$.
4. There is an identity morphism $\mathrm{id}_a : a \to a$ for each object $a \in \mathbf{Obj}$, such that for any other morphism $m : a \to a$, the identity $\mathrm{id}_a \circ m = m \circ \mathrm{id}_a = m$ holds.

A *subcategory* of a category **C** is a category **D** with $\mathbf{Obj}(\mathbf{D}) \subseteq \mathbf{Obj}(\mathbf{C})$ and $\mathbf{Mor}(\mathbf{D}) \subseteq \mathbf{Mor}(\mathbf{C})$.

Remark 4.1 The reader should note the choice of terms in the definition of a category. In particular, we have chosen the objects and morphisms to be *classes* rather than *sets*. Informally, these are collections of objects that share a particular property of interest. Importantly, all sets are classes. However, not all classes are sets. For instance, the class of all sets is not a set. This is a technical way to avoid Russell's paradox, which arises from the question "Does the set of all sets contain itself?"

Categories are a unifying feature of mathematics. For instance, Table 4.1 lists some elementary examples of categories (a few of which are defined later in this chapter).

Example 4.1 We can realize any group G as a category **G** with one object $*$. Define $\mathbf{Mor}(\mathbf{G}) = G$ and the composition of morphisms to simply be the multiplication of group elements. The axioms of a category are automatically satisfied by the axioms of the group.

Substantial inferential power arises from converting a problem in one category into another where the solution is easier. For instance, algebraic topology is the discipline of translating topological problems (in **Top**) into algebraic ones (such as **Grp** or **Vec**). The central theme of this book is that signal processing is a similar translation; from a category of signals into a more convenient one. For instance, the Signal Embedding Theorem 2.2 permits a translation of the collection of wireless signals into $\mathbf{Man}^{\mathbf{k}}$. In this chapter, we go further, translating the problem into **Cell**.

Any such translation of one category into another should preserve its structure. The precise description of that translation leads to the definition of a functor.

Definition 4.2 A *functor* **F** from a category **C** to another **D** assigns each object $X \in \mathbf{C}$ to an object $\mathbf{F}(X) \in \mathbf{D}$ and a morphism $\mathbf{F}(f) : \mathbf{F}(X) \to \mathbf{F}(Y) \in \mathbf{Mor}(\mathbf{D})$ to each morphism $f : X \to Y \in \mathbf{Mor}(\mathbf{C})$ according to the following rules:

Table 4.1 Several examples of frequently used categories

Category	Objects	Morphisms
Set	Sets	Functions
Open	Open sets	Inclusions
Top	Topological spaces	Continuous functions
Cell	Cell complexes	Cellular maps
$\mathbf{Man}^k$	C^k manifolds	C^k maps
Simp	Simplicial complexes	Simplicial maps
Grp	Groups	Homomorphisms
Vec	Vector spaces	Linear maps
ShX	Sheaves on X	Sheaf morphisms
Kom	Chain complexes	Chain morphisms

1. $\mathbf{F}(\mathrm{id}_X) = \mathrm{id}_{\mathbf{F}(X)}$, and
2. either $\mathbf{F}(f \circ g) = \mathbf{F}(f) \circ \mathbf{F}(g)$ for every pair of composable morphisms in $\mathbf{C}$ (in which case we say $\mathbf{F}$ is *covariant*) or $\mathbf{F}(f \circ g) = \mathbf{F}(g) \circ \mathbf{F}(f)$ (in which case we say that $\mathbf{F}$ is *contravariant*).

Without qualification, the term *functor* could mean either kind of functor.

Example 4.2 Group representations are functors. Specifically, consider the category $\mathbf{G}$ given in Example 4.1. A *representation* is a functor from $\mathbf{G}$ to **Vec**. For instance, the group of face moves of a Rubik's cube can be represented as a collection of 54×54 permutation matrices. In order to study the group of face moves of the Rubik's cube, it suffices to study these matrices, since the functor gives a perfect representation of the group.

Sometimes functors are *forgetful*; they lose information. The next example shows how information can be lost by a functor. This is sometimes desirable in applications, since forgetful functors can be substantially easier to compute or substantially more robust to variations.

Example 4.3 Consider the category $\mathbf{C}$ in which the objects are vector spaces and a morphism exists between $V, W \in \mathbf{C}$ if $V \subseteq W$. In this case, computing the dimension of a vector space is a forgetful covariant functor from $\mathbf{C}$ to the category of natural numbers in which relations like $a \leq b$ are the morphisms.

Example 4.4 One particularly important category for sheaves is the *face category* $\mathscr{F}X$ of a cell complex X, in which the objects are cells of X and attachments $a \rightsquigarrow b$ between cells a and b are the morphisms. *Sheaves are covariant functors from the face category into some other category* $\mathbf{C}$. For instance, a covariant functor from $\mathscr{F}X$ to **Vec** is a sheaf of vector spaces.

Example 4.5 In the category of sheaves, if $f : X \to Y$ is a cellular map, then the pushforward f_* and pullback f^* operations (Sect. 3.3.1) are functors between the categories of sheaves on X and Y.

Functors between a pair of categories can also be translated from one to another. This somewhat abstract idea is of considerable importance in the theory of sheaves, and is the reason for the prevalence of commutative diagrams in the theory. Almost every time a commutative diagram appears, it represents a *natural transformation of functors*.

Definition 4.3 Given a pair of covariant functors $\mathbf{F} : \mathbf{C} \to \mathbf{D}$ and $\mathbf{G} : \mathbf{C} \to \mathbf{D}$ from one category to another, a *natural transformation* $\mathbf{N}$ is an assignment of a morphism $\mathbf{N}_X : \mathbf{F}(X) \to \mathbf{G}(X)$ to each object $X, Y \in \mathbf{Ob}(\mathbf{C})$ such that for each morphism $(f : X \to Y) \in \mathbf{Mor}(\mathbf{C})$, the identity $\mathbf{N}_Y \circ \mathbf{F}(f) = \mathbf{G}(f) \circ \mathbf{N}_X$ holds. (We have that $\mathbf{N}_Y : \mathbf{F}(Y) \to \mathbf{G}(Y)$.) We represent this identity by saying that the following diagram *commutes*.

$$\begin{array}{ccc} F(X) & \xrightarrow{\mathbf{N}_X} & G(X) \\ \Big\downarrow{\mathbf{F}(f)} & & \Big\downarrow{\mathbf{G}(f)} \\ F(Y) & \xrightarrow{\mathbf{N}_Y} & G(Y) \end{array}$$

Exercise 4.1 Natural transformations can also be defined between

1. a pair of contravariant functors,
2. between contravariant and a covariant functor, and
3. between covariant and a contravariant functor.

Modify the definition to support these other cases.

One can think of natural transformations as being morphisms between functors. Therefore, we can define the category $\mathbf{Fun}(\mathbf{C}, \mathbf{D})$ whose objects are functors between categories $\mathbf{C}$ and $\mathbf{D}$ and morphisms are natural transformations. This is not as abstract as it might seem. Sheaf morphisms (Definition 3.4) are natural transformations of sheaves, considered as covariant functors, so the category of sheaves of $\mathbf{C}$ over a space X is $\mathbf{Fun}(\mathscr{F}X, \mathbf{C})$.

4.1.1 Detectors are Functors

A *signal* is a collection of related measurements. The topology on these measurements indicates how a given measurement responds to noise. A signal whose measurements are taken from a discrete set has a rather brittle response to noise. Either it is completely robust (and is not changed by noise) or its value is perturbed rather far from the correct one. On the other hand, a signal whose measurements are taken from a manifold can depend more smoothly on perturbations. A *detector* strips the topological structure from signals; often translating smoothly-varying signals to ones that have been quantized.

Definition 4.4 A *detector* is a functor to a subcategory of **Set**.

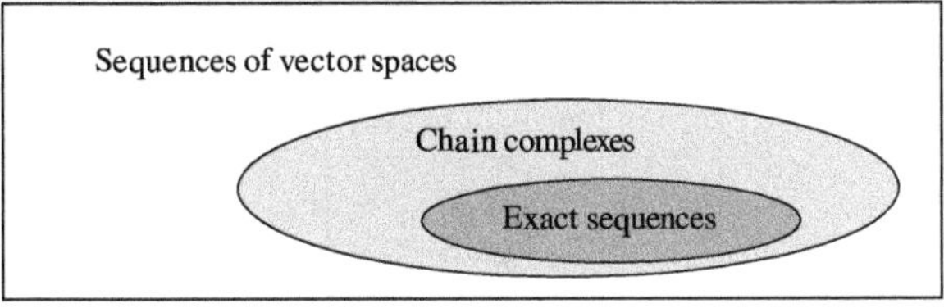

Fig. 4.1 Several types of sequences of vector spaces that play a role in the study of sheaves

This definition is somewhat subtle, as it encodes the "intent" of the filter's designer. In particular, every topological space is also a set. However, by specifically indicating that the output of a detector is a set and *not* a topological space, we indicate that continuity is inappropriate when discussing the output.

Example 4.6 (compare with Example 3.25) A threshold detector can be realized as a functor. Specifically, a threshold detector takes a continuous function $f \in C(\mathbb{R})$ and returns the open set on which $f(x) > T$ for some threshold $T \in \mathbb{R}$. We specify the input to this functor to be the category **C** in which each continuous function is an object. Let the morphisms of **C** be given by $f \to g$ whenever $f(x) > g(x)$ for all $x \in \mathbb{R}$. The detector is then given by a morphism $\mathbf{F} : \mathbf{C} \to$ **Open** that assigns $F(f) = \{x \in \mathbb{R} : f(x) > T\}$ to each $f \in C(\mathbb{R})$. Observe that if $f \to g$, then $F(g) \subset F(f)$. Hence F is a contravariant functor.

This simple example is the first of many more powerful detectors. We will spend the rest of this chapter examining detectors that arise from the *cohomology* of sheaves.

4.2 Exact Sequences

In order to develop the theory of *cohomology*, we need to first develop a mechanism for managing related information in collections of vector spaces. In doing so, we need to study sequences of vector spaces, with maps from one to the next. There are several different kinds of sequences that play a role in our analysis, as indicated by Fig. 4.1. In many situations, the most interesting ones are the non-exact chain complexes.

Definition 4.5 A *sequence* of vector spaces $(V_\bullet, d_\bullet)$ is a collection of vector spaces V_i and linear maps d_i between them as follows

$$\cdots \xrightarrow{d_{i-1}} V_i \xrightarrow{d_i} V_{i+1} \xrightarrow{d_{i+1}} \cdots$$

Each V_i is called a *term* of the sequence.

A sequence of vector spaces is *exact* when $\ker d_i = \text{image}\, d_{i-1}$. Exact sequences are like telescoping series in that most terms "cancel," as shown schematically in

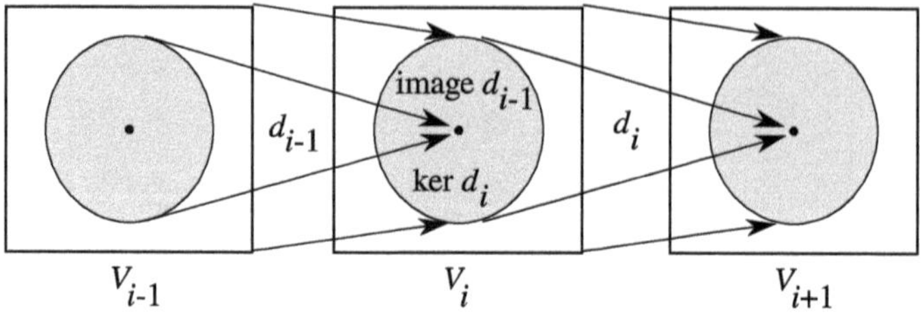

Fig. 4.2 A schematic of an exact sequence of vector spaces

Fig. 4.2. We call an exact sequence with finitely many nontrivial spaces, such as

$$0 \to A \to B \to C \to 0,$$

a *short exact sequence*.

Exercise 4.2 Show that the following sequence of vector spaces with the linear maps as given is exact.

$$0 \xrightarrow{\begin{pmatrix}0\\0\end{pmatrix}} \mathbb{R}^2 \xrightarrow{\begin{pmatrix}1&0\\0&0\\0&1\\0&0\end{pmatrix}} \mathbb{R}^4 \xrightarrow{\begin{pmatrix}0&1&0&0\\0&0&0&1\\0&1&0&1\end{pmatrix}} \mathbb{R}^3 \xrightarrow{\begin{pmatrix}1&1&-1\end{pmatrix}} \mathbb{R} \xrightarrow{0} 0$$

Exercise 4.3 (Useful properties of short exact sequences)

1. Show that if the following sequence is exact, then f is injective:

$$0 \longrightarrow V \xrightarrow{f} W.$$

2. Show that if the following sequence is exact, then f is surjective:

$$V \xrightarrow{f} W \longrightarrow 0.$$

3. Show that if the following sequence is exact, then f is an isomorphism:

$$0 \longrightarrow V \xrightarrow{f} W \longrightarrow 0.$$

Exact sequences are important because they generalize the dimension theorem (or rank nullity theorem) from basic linear algebra. Recall that the dimension theorem states that if $f : V \to W$ is a linear map between finite dimensional vector spaces, then

$$\dim V = \dim \ker f + \operatorname{rank} f, \text{ and } \dim W = \dim \operatorname{coker} f + \operatorname{rank} f,$$

where we recall that coker f is the orthogonal complement of the image of f. Observe that if

$$0 \longrightarrow A \longrightarrow V \xrightarrow{f} W \longrightarrow B \longrightarrow 0$$

is exact, then $A \cong \ker f$ and $B \cong \operatorname{coker} f$. Hence,

$$V \cong A \oplus \text{image } f \cong \ker f \oplus \text{image } f$$

and

$$W \cong B \oplus \text{image } f \cong \operatorname{coker} f \oplus \text{image } f,$$

both of which are equivalent to the dimension theorem.

We will make extensive use of a particular sequence of vector spaces associated with a sheaf over cell complexes, which is typically not exact. Instead, such sequences are "approximately" exact. Due to their historical association with cell complexes (and particularly simplicial complexes), this kind of sequence is called a *chain complex*.

Definition 4.6 A *chain complex* is a sequence of vector spaces

$$\cdots \xrightarrow{d_{i-1}} V_i \xrightarrow{d_i} V_{i+1} \xrightarrow{d_{i+1}} \cdots$$

in which $d_{i+1} \circ d_i = 0$ for all i.

Observe that if $(V_\bullet, d_\bullet)$ is a chain complex, then image $d_{i-1} \subseteq \ker d_i$. Therefore, exact sequences are automatically chain complexes. But as the following exercise shows, the converse is not true.

Exercise 4.4 Show that the following sequence is a chain complex, but not an exact sequence.

$$0 \longrightarrow \mathbb{R}^4 \xrightarrow{d^0} \mathbb{R}^6 \xrightarrow{d^1} \mathbb{R}^4 \xrightarrow{d^2} \mathbb{R} \longrightarrow 0.$$

where

$$d^0 = \begin{pmatrix} 1 & -1 & 0 & 0 \\ 1 & 0 & -1 & 0 \\ 1 & 0 & 0 & -1 \\ 0 & 1 & -1 & 0 \\ 0 & 1 & 0 & -1 \\ 0 & 0 & 1 & -1 \end{pmatrix},$$

$$d^1 = \begin{pmatrix} 0 & 0 & 0 & 1 & -1 & 1 \\ 0 & -1 & 1 & 0 & 0 & -1 \\ 1 & 0 & -1 & 0 & 1 & 0 \\ -1 & 1 & 0 & -1 & 0 & 0 \end{pmatrix},$$

and

$$d^2 = \begin{pmatrix} 1 & 1 & 1 & 1 \end{pmatrix}.$$

The perspective of category theory is that the best way to study objects is to also study the morphisms between them. In the case of **Vec**, one studies vector spaces by the linear maps between them. A natural direction of study for chain complexes, then, is to look at a collection of term-by-term linear maps. In this case, we are led to the following diagram from two chain complexes $(V_\bullet, d_\bullet)$ and $(W_\bullet, e_\bullet)$

$$\begin{array}{ccccccc} \cdots & \xrightarrow{d_{i-1}} & V_i & \xrightarrow{d_i} & V_{i+1} & \xrightarrow{d_{i+1}} & \cdots \\ & & \downarrow{\scriptstyle L_i} & & \downarrow{\scriptstyle L_{i+1}} & & \\ \cdots & \xrightarrow{e_{i-1}} & W_i & \xrightarrow{e_i} & W_{i+1} & \xrightarrow{e_{i+1}} & \cdots \end{array} \tag{4.1}$$

where L_i are linear.

Definition 4.7 A collection of linear maps $L_i : V_i \to W_i$ is a *chain map* from the chain complex $(V_\bullet, d_\bullet)$ to $(W_\bullet, e_\bullet)$ if the diagram (4.1) commutes. These are the morphisms in the category **Kom** of chain complexes of vector spaces.

Example 4.7 The diagram below exhibits a chain map (columns) between two chain complexes (rows)

$$\begin{array}{ccccccccc} 0 & \longrightarrow & \mathbb{R}^2 & \xrightarrow{\begin{pmatrix} 1 & -1 \\ -1 & 1 \end{pmatrix}} & \mathbb{R}^2 & \xrightarrow{\begin{pmatrix} 1 & 1 \\ 1 & 1 \end{pmatrix}} & \mathbb{R}^2 & \longrightarrow & 0 \\ & & \downarrow{\scriptstyle \mathrm{id}} & & \downarrow{\scriptstyle \mathrm{id}} & & \downarrow & & \\ 0 & \longrightarrow & \mathbb{R}^2 & \xrightarrow[\begin{pmatrix} 1 & -1 \\ -1 & 1 \end{pmatrix}]{} & \mathbb{R}^2 & \longrightarrow & 0 & \longrightarrow & 0 \end{array}$$

Observe that each square commutes. In contrast, the similar looking diagram

$$
\begin{array}{ccccccccc}
0 & \longrightarrow & \mathbb{R}^2 & \xrightarrow{\begin{pmatrix} 1 & -1 \\ -1 & 1 \end{pmatrix}} & \mathbb{R}^2 & \longrightarrow & 0 & \longrightarrow & 0 \\
 & & \Big\downarrow \text{id} & & \Big\downarrow \text{id} & & \Big\downarrow & & \\
0 & \longrightarrow & \mathbb{R}^2 & \xrightarrow[\begin{pmatrix} 1 & -1 \\ -1 & 1 \end{pmatrix}]{} & \mathbb{R}^2 & \xrightarrow[\begin{pmatrix} 1 & 1 \\ 1 & 1 \end{pmatrix}]{} & \mathbb{R}^2 & \longrightarrow & 0
\end{array}
$$

is *not* a chain map.

Let us return to the difference between chain complexes and exact sequences. The terms in chain complexes can have larger kernels than in an exact sequence. The maps collapse a larger set than simply the elements of the image of the previous map.

Definition 4.8 The *homology* of a chain complex $(V_\bullet, d_\bullet)$ is the collection of vector spaces $H_\bullet$ given by

$$H_i(V_\bullet) = \ker d_i / \text{image}\, d_{i-1}.$$

The homology spaces are all zero if and only if a chain complex is an exact sequence.

Example 4.8 The homology of the sequence

$$0 \longrightarrow \mathbb{R}^2 \xrightarrow[d_0]{\begin{pmatrix} 1 & -1 \\ -1 & 1 \end{pmatrix}} \mathbb{R}^2 \xrightarrow[d_1]{\begin{pmatrix} 1 & 1 \\ 1 & 1 \end{pmatrix}} \mathbb{R}^2_{d_2} \longrightarrow 0$$

is given by the sequence of spaces $\mathbb{R}, 0, \mathbb{R}, 0, 0, \ldots$. Explicitly,

$$\ker d_0 = \ker \begin{pmatrix} 1 & -1 \\ -1 & 1 \end{pmatrix} = \text{span}\left\{\begin{pmatrix} 1 \\ 1 \end{pmatrix}\right\},$$

$$\ker d_1 = \ker \begin{pmatrix} 1 & 1 \\ 1 & 1 \end{pmatrix} = \text{span}\left\{\begin{pmatrix} 1 \\ -1 \end{pmatrix}\right\} = \text{image} \begin{pmatrix} 1 & -1 \\ -1 & 1 \end{pmatrix} = \text{image}\, d_0.$$

In contrast, the homology of

$$0 \longrightarrow \mathbb{R}^2 \xrightarrow[d_0]{\begin{pmatrix} 1 & -1 \\ -1 & 1 \end{pmatrix}} \mathbb{R}^2 \xrightarrow[d_1]{(1\ 1)} \mathbb{R} \xrightarrow[d_2]{} 0$$

is given by the sequence of spaces $\mathbb{R}, 0, 0, \ldots$, since the matrix $\begin{pmatrix}1 & 1\end{pmatrix}$ represents a surjective linear map.

Proposition 4.1 *Chain maps induce linear maps between the associated homology spaces. Therefore, each H_i is a functor* **Kom** $\to$ **Vec**.

Proof Suppose that L_i is a chain map from $(V_\bullet, d_\bullet)$ to $(W_\bullet, e_\bullet)$, two chain complexes. If $x \in \ker d_i$, then $d_i x = 0$. Then $e_i \circ L_i x = L_{i+1} \circ d_i x = L_{i+1} 0 = 0$ according to (4.1). Hence $L_i x \in \ker e_i$. Similarly, if $x \in \operatorname{image} d_{i-1}$, then there is a $y \in V_{i-1}$ for which $d_{i-1} y = x$. Again, because L_i is a chain map, $e_{i-1} \circ L_{i-1} y = L_i \circ d_{i-1} y = L_i x$, so that $L_i x \in \operatorname{image} e_{i-1}$. Therefore $L_\bullet$ descends to a map on each quotient H_i. □

If we have a short exact sequence $0 \to A_i \to B_i \to C_i \to 0$ of chain complexes in which the arrows represent chain maps, this induces linear maps between the homologies $H_i(A) \to H_i(B) \to H_i(C)$. But there something more substantial:

Lemma 4.1 *(The snake lemma) There is a* connecting map $\delta_i : H_i(C) \to H_{i+1}(A)$ *which makes the following sequence exact*

$$\cdots \longrightarrow H_i(A) \longrightarrow H_i(B) \longrightarrow H_i(C) \xrightarrow{\delta_i} H_{i+1}(A) \longrightarrow \cdots$$

which we call the long exact sequence *associated to the short exact sequence $0 \to A_i \to B_i \to C_i \to 0$ of chain complexes.*

By hypothesis, we have the following commutative diagram

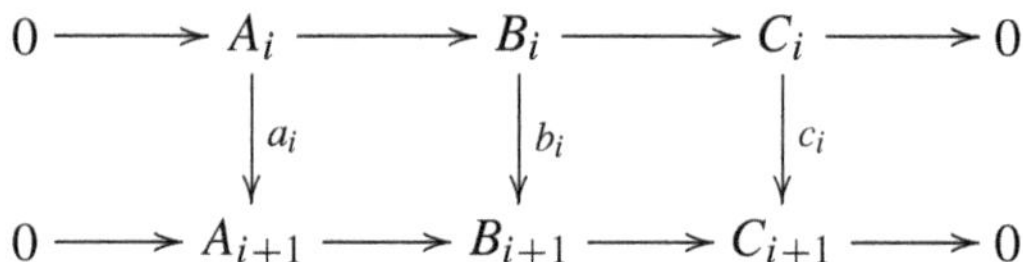

in which the rows are exact sequences. The snake lemma shows how to get from the upper right part of the diagram to the lower left. Everything else in the long exact sequence already follows via functoriality (Proposition 4.1). Building one instance of the connecting map extends the long exact sequence by three terms, by "snaking" them onto the end.

Proof One performs the following *diagram chase*, beginning in the upper right corner with $x \in \ker c_i \subseteq C_i$.

1. Since the map $B_i \to C_i$ is surjective, this means that there is a $y \in B_i$ whose image is x.
2. This y corresponds to a $\tilde{y} \in B_{i+1}$,
3. which by commutativity of the diagram, $\tilde{y} \in \ker(B_{i+1} \to C_{i+1})$.
4. By exactness, there must be a $\tilde{x} \in A_{i+1}$ whose image is $\tilde{y}$.

Therefore, let $\delta_i : \ker c_i \to A_{i+1}$ be given by $x \mapsto \tilde{x}$. Because the diagram commutes, δ_{i+1} takes the image of x to the image of $\tilde{i}$. Therefore, δ_i maps the kernel of c_i to the kernel of a_{i+1} and the image of c_{i-1} to the image of a_i. Therefore, δ is actually a chain map, and so it induces a map on homology $H_i(C) \to H_{i+1}(A)$ and makes the resulting sequence exact. □

4.3 Sheaf Cohomology

Sheaves of abelian groups give rise to a special chain complex that additionally captures the topology and combinatorics of the underlying cell complex. We follow Shepard (1980) closely, and study the homology of this complex, called *sheaf cohomology*, which captures certain useful properties and is a powerful detector on sheaves.

4.3.1 Orientation

Orientation of cells in a cell complex determines how information stored in a sheaf is tested for consistency. Suppose that $\mathscr{S}$ is a sheaf on a cell complex in which v and w are attached to an edge e, as shown in the diagram below

$$\mathscr{S}(v) \xrightarrow{\mathscr{S}(v \rightsquigarrow e)} \mathscr{S}(e) \xleftarrow[\mathscr{S}(w \rightsquigarrow e)]{} \mathscr{S}(w).$$

In order to extend a section s that is supported on v and w to e, we must have that

$$(\mathscr{S}(v \rightsquigarrow e))\, s(v) = (\mathscr{S}(w \rightsquigarrow e))\, s(w) = s(e).$$

If $\mathscr{S}$ is a sheaf of abelian groups, this equation can be written

$$+\,(\mathscr{S}(v \rightsquigarrow e))\, s(v) - (\mathscr{S}(w \rightsquigarrow e))\, s(w) = 0$$

or

$$-\,(\mathscr{S}(v \rightsquigarrow e))\, s(v) + (\mathscr{S}(w \rightsquigarrow e))\, s(w) = 0.$$

This suggests that $v \rightsquigarrow e$ and $w \rightsquigarrow e$ should be assigned opposite signs. The concept of *orientation* makes this assignment precise.

We begin by defining the orientation of a nonsingular linear map $L : V \to V$ between two vector spaces V.

Definition 4.9 The *orientation* of a nonsingular linear map $L : V \to V$ is the sign of its determinant.

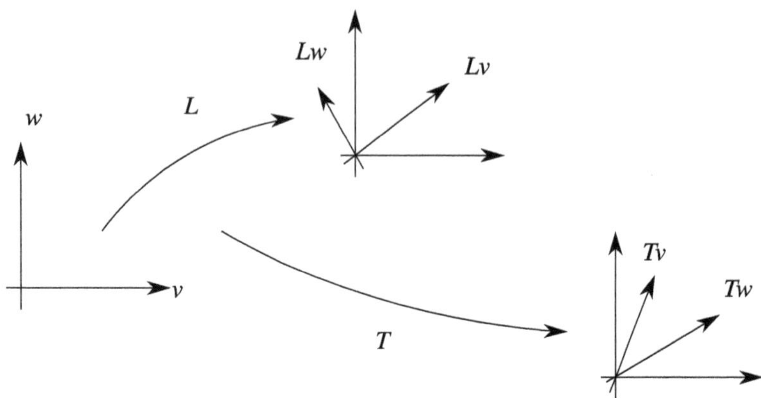

Fig. 4.3 The orientation of a two linear maps $L, T : \mathbb{R}^2 \to \mathbb{R}^2$. The orientation of L is $+1$, while the orientation of T is -1

This somewhat opaque definition has a convenient interpretation. Consider the identity matrix $\text{id} : V \to V$, which has determinant $+1$. Switching two rows or columns of this matrix results in a matrix whose determinant has opposite sign. This takes a basis $a_1, \ldots, a_n$ for V to one in which the basis elements themselves are the same, but are listed in a different order, as shown in Fig. 4.3.

Definition 4.10 Suppose that $L : V \to V$ is a nonsingular linear map that takes a basis $a_1, \ldots, a_n$ to another basis $b_1, \ldots, b_n$. If the orientation of L is $+1$, we say that the orientations of the two bases *agree*. If the orientation of L is -1, then we say that the orientations of the bases *disagree*.

Suppose that $a_1, \ldots, a_n$ is an orthonormal basis for $\mathbb{R}^n$ and that D^n is a disk in $\mathbb{R}^n$. Consider n smooth maps $b_i : D^n \to \mathbb{R}^n$ chosen so that for each $x \in D^n$ the set $\{b_1(x), \ldots, b_n(x)\}$ is an orthonormal basis for $\mathbb{R}^n$. Because of this choice, there is a unique linear map $T(x) : \mathbb{R}^n \to \mathbb{R}^n$ that transforms the basis $a_1, \ldots, a_n$ to the basis $b_1(x), \ldots, b_n(x)$. The determinant of this T is ± 1, since it is an isomorphism. Observe that the function defined by

$$\det \left(T(x) \begin{pmatrix} b_1(x) \\ \vdots \\ b_n(x) \end{pmatrix} \right)$$

is constant, since it is continuous, takes either the value $+1$ or -1, and is defined on a connected set. This allows us to compare whether the orientation of the vector space $\mathbb{R}^n$ agrees with the orientation of D^n.

Definition 4.11 An *orientation* of a disk D^n is a choice of n smooth maps $b_i : D^n \to \mathbb{R}^n$ called the *orientation frame* (or *orientation*), chosen so that for each $x \in D^n$ the set $\{b_1(x), \ldots, b_n(x)\}$ is a basis for $\mathbb{R}^n$. By the previous discussion, the

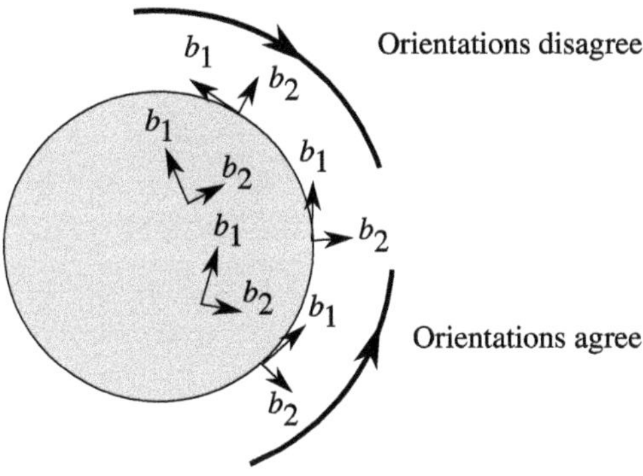

Fig. 4.4 Two oriented 1-cells attached to an oriented 2-cell

orientation frame can be chosen up to agreement of orientations. A different choice of orientation frame which agrees with $b_1, \ldots, b_n$ is said to have the same orientation.

Suppose that $A \rightsquigarrow B$ is the attachment of an n-dimensional cell A to the boundary of an $n+1$-dimensional cell B, via an attaching map ϕ that is a diffeomorphism. Let $C \subset \partial B$ be the portion of B that $\phi(C) = A$. Suppose that an orientation $a_1, \ldots, a_n$ is chosen for A and an orientation $b_1, \ldots, b_{n+1}$ is chosen for B in which b_{n+1} is normal to the boundary of B. Since ϕ was assumed to be a diffeomorphism, then the derivative $d\phi$ is a linear isomorphism. Therefore, $\{d\phi(b_1), \ldots, d\phi(b_n)\}$ is an orientation on A, called the *induced orientation* (see Fig. 4.4). We define the *index* of $A \rightsquigarrow B$ to be

$$[A : B] = \begin{cases} 1 & \text{if the orientation of } A \text{ agrees with the one induced by } B \\ -1 & \text{if the orientation of } A \text{ disagrees with the one induced by } B \end{cases}$$

Conventionally, we define $[A : B] = 0$ if A is not attached to B. This has a particularly simple form in the case of abstract simplicial complexes if σ and τ are both simplices whose dimension differs by 1. If $\tau = (t_1, \ldots, t_n)$ and $\sigma = (t_1, \ldots, t_{k-1}, t_{k+1}, \ldots, t_n)$, then $[\sigma : \tau] = (-1)^k$.

4.3.2 Definition of Sheaf Cohomology

Suppose that $\mathscr{S}$ is a sheaf of abelian groups on a cell complex X. The *k-th cochain group with compact supports* $C_c^k(X; \mathscr{S})$ of $\mathscr{S}$ is the direct sum of the stalks over the k-cells of X. Explicitly, they are given by

$$C_c^k(X; \mathscr{S}) = \bigoplus_{\sigma \in X^k} \mathscr{S}(\sigma).$$

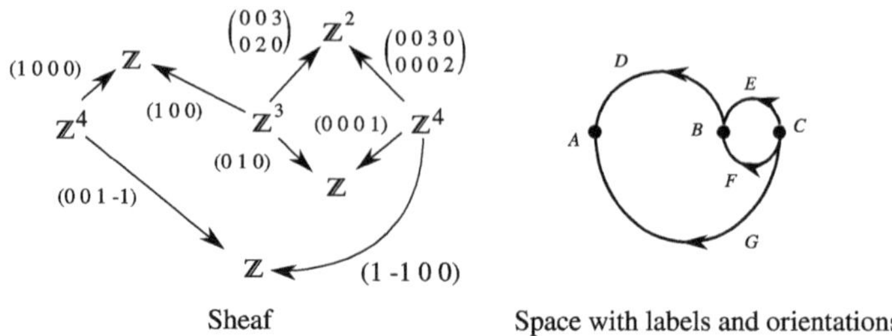

Fig. 4.5 The sheaf in Example 4.9 (*left*) and its base space (*right*)

Notice that elements of $C_c^k(X;\mathscr{S})$ are functions from the k-cells to the stalks over those cells. Most of the time, it is useful to exclude those cells whose closures are not compact, since these cells represent external connections. We therefore define the *k-th cochain group*

$$C^k(X;\mathscr{S}) = \bigoplus_{\sigma\in X^k:\overline{\sigma} \text{ is compact}} \mathscr{S}(\sigma).$$

The *k-th coboundary homomorphism* (or *coboundary map*) is the homomorphism $d^k : C_c^k(X;\mathscr{S}) \to C_c^{k+1}(X;\mathscr{S})$ given by

$$(d^k f)(\tau) = \sum_{\sigma\in X^k} [\sigma:\tau]\mathscr{S}(\sigma \rightsquigarrow \tau) f(\sigma)$$

where $\tau \in X^{k+1}$ and $f \in C_c^k(X;\mathscr{S})$. We define d^k on $C^k(X;\mathscr{S})$ with the same formula, but restrict our attention to the cells with compact closures.

Example 4.9 Consider the sheaf given by the diagram in Fig. 4.5. Using the orientations indicated on the figure, the coboundary map d^0 has the matrix form

$$d^0 = \left(\begin{array}{cccc|ccc|cccc}
1&0&0&0&-1&0&0&0&0&0&0\\ \hline
0&0&0&0&0&0&3&0&0&-3&0\\ \hline
0&0&0&0&0&2&0&0&0&0&-2\\ \hline
0&0&0&0&0&1&0&0&0&0&-1\\ \hline
0&0&1&-1&0&0&0&-1&1&0&0
\end{array}\right),$$

where the blocks of columns correspond to A, B, and C respectively and the blocks of rows correspond to D, E, F, and G respectively.

Lemma 4.2 *If $\mathscr{S}$ is a sheaf on a cell complex X, then the composition $d^{k+1}\circ d^k : C^k(X;\mathscr{S}) \to C^{k+2}(X;\mathscr{S})$ is the zero map, so that $(C^\bullet(X;\mathscr{S}), d^\bullet)$ is a chain complex.*

Proof (following Shepard (1980) closely) We begin by observing that if $\sigma \rightsquigarrow \gamma \rightsquigarrow \tau$ and each attachment increases the dimension by one, then there are exactly two choices for γ. This follows immediately from the Jordan curve theorem,[1] which states that the complement of S^k in S^{k+1} has two connected components. In our case, the image of the attaching map from τ to γ is a sphere of dimension 1 less than that of τ; the image of the attaching map from τ to σ is a sphere of dimension 2 less than that of τ. (This is false if X is a general CW complex where attaching maps may not be embeddings, since the image of the attaching map might not be a sphere of the proper dimension.) We therefore have the attachment diagram

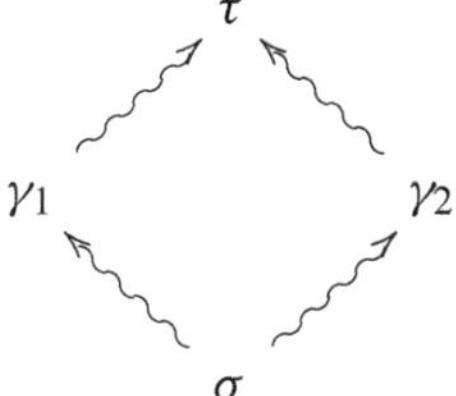

These two choices γ_1, γ_2 satisfy $[\sigma : \gamma_1][\gamma_1 : \tau] + [\sigma : \gamma_2][\gamma_2 : \tau] = 0$. We can prove this identity using abstract simplicial complexes without loss of generality, since it is a local statement depending only on the four cells under discussion. Suppose that $\sigma = (s_1, \ldots, s_n)$ and $\tau = (s_1, \ldots, s_{n+2})$. Then $\gamma_1 = (s_1, \ldots, s_n, s_{n+1})$ and $\gamma_2 = (s_1, \ldots, s_n, s_{n+2})$. Evidently $[\sigma : \gamma_1] = [\sigma : \gamma_2]$, but $[\gamma_1 : \tau]$ and $[\gamma_2 : \tau]$ are of opposite signs.

We can then treat the calculation of the composition directly:

$$
\begin{aligned}
\left(d^{k+1} \circ d^k f\right)(\tau) &= \sum_{\gamma \in X^{k+1}} [\gamma : \tau] \mathscr{S}(\gamma \rightsquigarrow \tau)(d^k f)(\gamma) \\
&= \sum_{\gamma \in X^{k+1}} [\gamma : \tau] \mathscr{S}(\gamma \rightsquigarrow \tau) \left(\sum_{\sigma \in X^k} [\sigma : \gamma] \mathscr{S}(\sigma \rightsquigarrow \gamma) f(\sigma) \right) \\
&= \sum_{\gamma \in X^{k+1}} \sum_{\sigma \in X^k} [\sigma : \gamma][\gamma : \tau] \mathscr{S}(\sigma \rightsquigarrow \tau) f(\sigma) \\
&= \sum_{\sigma \in X^k} ([\sigma : \gamma_1][\gamma_1 : \tau] + [\sigma : \gamma_2][\gamma_2 : \tau]) \mathscr{S}(\sigma \rightsquigarrow \tau) f(\sigma) \\
&= 0.
\end{aligned}
$$

□

[1] The proof of the Jordan curve theorem (for instance see Hatcher (2002, Sect. 2.B)) usually relies on singular homology, which does not depend on the sheaf cohomology theory we are developing here.

Definition 4.12 The *cohomology* $H^\bullet(X; \mathscr{S})$ of a sheaf $\mathscr{S}$ of abelian groups on a cell complex X is the homology of the chain complex $(C^\bullet(X; \mathscr{S}), d^\bullet)$. When the base space is implied from context, we will sometimes write $H^\bullet(\mathscr{S})$ for the cohomology of $\mathscr{S}$. Similarly, the *cohomology of* $\mathscr{S}$ *with compact supports* $H_c^\bullet(X; \mathscr{S})$ is the homology of the chain complex $(C_c^\bullet(X; \mathscr{S}), d^\bullet)$.

Remark 4.2 We call $H^\bullet(X; \mathscr{S})$ "cohomology" as opposed to "homology," because in the context of cell complexes the term "homology" refers to a chain complex in which the maps decrease cell dimension. According to the definition of the coboundary map, d^k converts stalks over cells of dimension k to $k+1$, which is the opposite.

Example 4.10 Consider the following directed graph X, which we interpret as a cell complex

$$\longrightarrow \bullet \rightrightarrows \bullet \longrightarrow$$

The flow sheaf $\mathscr{F}$ (Example 3.6) on this space is given by

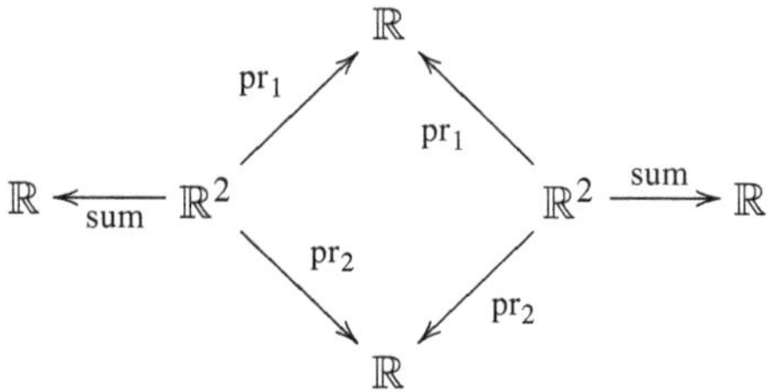

The cochain complex for this sheaf is given by

$$0 \longrightarrow \mathbb{R}^4 \xrightarrow{d} \mathbb{R}^2 \longrightarrow 0$$

where the coboundary map has the structure

$$d = \begin{pmatrix} \mathrm{pr}_1 & -\mathrm{pr}_1 \\ \mathrm{pr}_2 & -\mathrm{pr}_2 \end{pmatrix} = \begin{pmatrix} 1 & 0 & -1 & 0 \\ 0 & 1 & 0 & -1 \end{pmatrix}.$$

The signs in d come from an assumed arbitrary choice of orientation on the cells. Since the cohomology is a vector space, the precise choice of orientations is irrelevant to the resulting cohomology since any consistent choice will result in isomorphic vector spaces. The resulting cohomology spaces are given by

$$H^k(X; \mathscr{F}) = \begin{cases} \mathbb{R}^2 & \text{if } k = 0 \\ 0 & \text{otherwise} \end{cases}$$

The cochain complex with compact supports for $\mathscr{F}$ is different, since there are two edges with non-compact closures, so

$$0 \longrightarrow \mathbb{R}^4 \xrightarrow{d=\begin{pmatrix} 1 & 0 & -1 & 0 \\ 0 & 1 & 0 & -1 \\ 1 & 1 & 0 & 0 \\ 0 & 0 & 1 & 1 \end{pmatrix}} \mathbb{R}^4 \longrightarrow 0$$

Since the matrix d has rank 3, this results in the compactly supported cohomology

$$H_c^k(X; \mathscr{F}) = \begin{cases} \mathbb{R} & \text{if } k = 0, 1 \\ 0 & \text{otherwise} \end{cases}$$

One could also interpret the compactly supported cohomology as being the cohomology of a somewhat different sheaf, namely

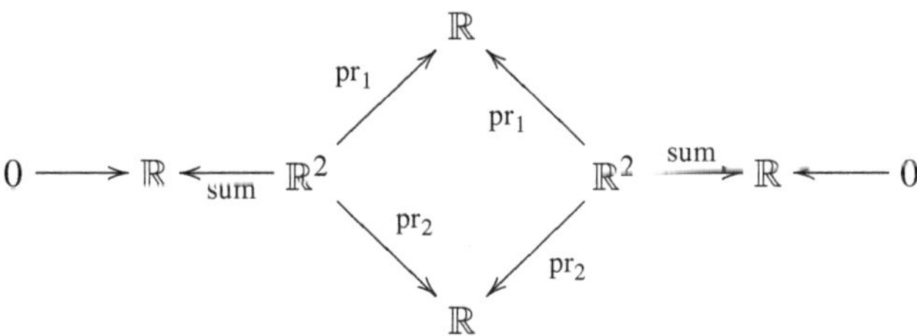

Theorem 4.1 *Sheaf morphisms induce chain maps between the cochain complexes of sheaves. Therefore sheaf cohomology is both a functor and a detector on sheaves.*

Proof Let $m : \mathscr{S} \to \mathscr{R}$ be a morphism of sheaves over a cellular map $f : X \to Y$. Then m extends linearly to a sequence of maps $m^k : C^k(Y; \mathscr{S}) \to C^k(X; \mathscr{R})$. Then for $\sigma \in X^k$ and $\tau \in X^{k+1}$ with $\sigma \rightsquigarrow \tau$, we have the following commutative diagram

$$\begin{array}{ccc} \mathscr{S}(f(\sigma)) & \xrightarrow{m_\sigma^k} & \mathscr{R}(\sigma) \\ {\scriptstyle \mathscr{S}(f(\sigma)\rightsquigarrow f(\tau))}\downarrow & & \downarrow{\scriptstyle \mathscr{R}(\sigma\rightsquigarrow\tau)} \\ \mathscr{S}(f(\tau)) & \xrightarrow{m_\tau^{k+1}} & \mathscr{R}(\tau) \end{array}$$

This leads to the following calculation with $s \in C^k(Y; \mathscr{S})$

$$\begin{aligned} \left((m^{k+1} \circ d^k)s\right)(\tau) &= \left(m_\tau^{k+1} \circ d_Y^k s\right)(\tau) \\ &= m_\tau^{k+1}\left(\sum_{f(\sigma)\in Y^k} [f(\sigma) : f(\tau)]\mathscr{S}(f(\sigma) \rightsquigarrow f(\tau))s(f(\sigma))\right) \end{aligned}$$

$$= \sum_{f(\sigma)\in Y^k} [f(\sigma) : f(\tau)] \left(m_\tau^{k+1} \circ \mathscr{S}(f(\sigma) \rightsquigarrow f(\tau))\right) s(f(\sigma))$$

$$= \sum_{\sigma\in X^k} [\sigma : \tau] \left(\mathscr{R}(\sigma \rightsquigarrow \tau) \circ m_\sigma^k\right) s(f(\sigma))$$

$$= \sum_{\sigma\in X^k} [\sigma : \tau] \mathscr{R}(\sigma \rightsquigarrow \tau) \left(m_\sigma^k s(f(\sigma))\right)$$

$$= \left((d_X^k \circ m^k)s\right)(\tau)$$

as desired. □

Example 4.11 Consider the cellular map of two directed graphs, given by the dashed arrows in the diagram

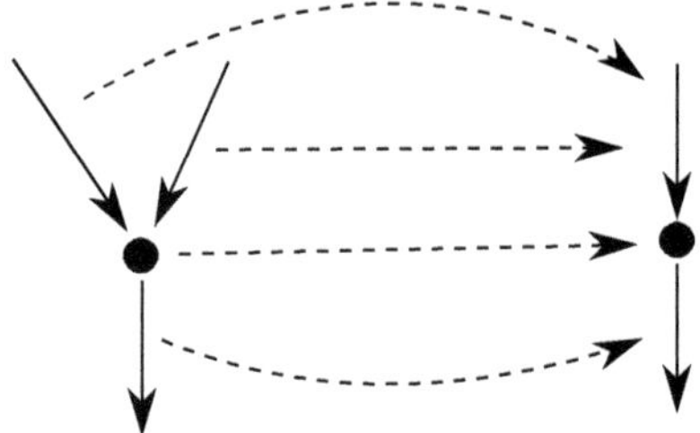

Recall that $\mathbb{F}_2$ is the field with two elements, which models binary arithmetic (addition = exclusive OR, multiplication = AND), and $\mathbb{F}_2^2$ is the 2-dimensional vector space over $\mathbb{F}_2$. The following diagram shows a sheaf morphism over the cellular map

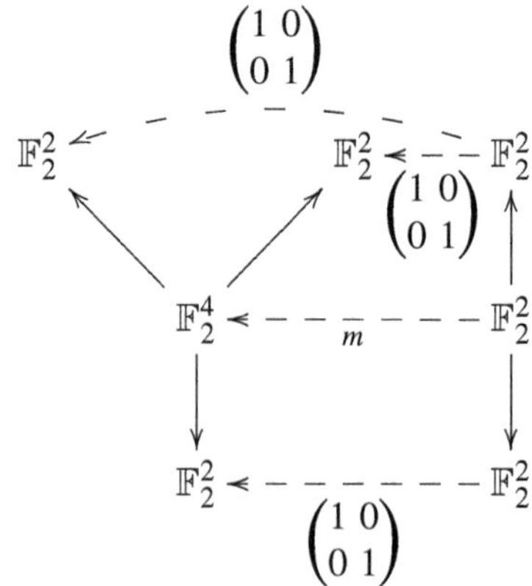

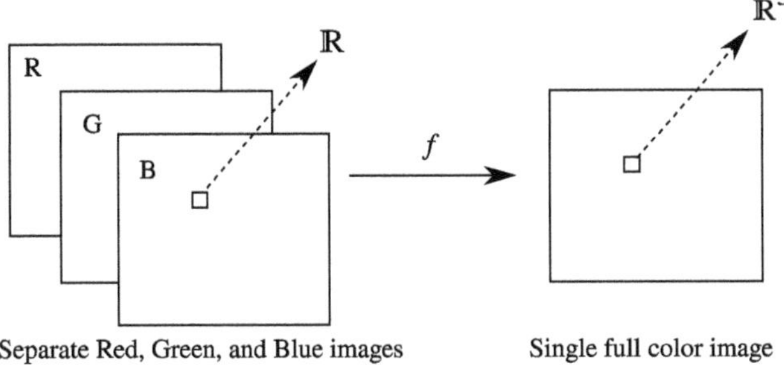

Fig. 4.6 Combining images representing different color channels into a single full-color image

where m is either $\begin{pmatrix} 1 & 0 \\ 0 & 0 \\ 0 & 0 \\ 0 & 1 \end{pmatrix}$ or $\begin{pmatrix} 0 & 1 \\ 1 & 1 \\ 1 & 1 \\ 1 & 0 \end{pmatrix}$. Observe that each square in the diagram above commutes, and that the arrows are reversed with respect to the cellular map. Both of these are defining features of a sheaf morphism.

This morphism induces a homomorphism on the sheaf cohomologies, which up to selection of bases for H^0 is given by the matrix for m.

The definition of the pushforward operation on sheaves induces a sheaf morphism, and therefore a homomorphism on cohomology. Under a certain important circumstance, this is an isomorphism.

Theorem 4.2 *(Vietoris mapping theorem (Bredon 1997, Section II.11)) Suppose that $f : X \to Y$ is a cellular map and that $\mathscr{S}$ is a sheaf on X. There is an induced map on cohomology $H^k(Y; f_*\mathscr{S}) \to H^k(X; \mathscr{S})$ by Proposition 4.1. If $H^k(f^{-1}(star\, y); \mathscr{S}) = 0$ for all cells $y \in Y$ and all $k > 0$, then this map is an isomorphism.*

The interested reader should consult Chapter II of Bredon (1997), which provides a self-contained exposition leading up to the proof of this theorem about half way through.

Example 4.12 Combining the red, green, and blue color channels to form a full-color image is a standard procedure in digital photography. Sometimes it is more convenient to work with separate color channels, and sometimes it is more convenient to work with the image as a whole. Exercise 3.1 called for the construction of the space of all images as a sheaf $\mathscr{I}$ over a grid $\mathbb{Z}^2$ of pixels. We represent the collection of separate color channels as a sheaf $\mathscr{C}$ over the disjoint union $\mathbb{Z}^2 \sqcup \mathbb{Z}^2 \sqcup \mathbb{Z}^2$, in which the restriction to each component (color channel) is a copy of $\mathscr{I}$.

We represent the process of combining separate color channels as a map f which takes each pixel location in a color channel to its corresponding location in the combined image, as shown in Fig. 4.6. The Vietoris mapping theorem applies to this

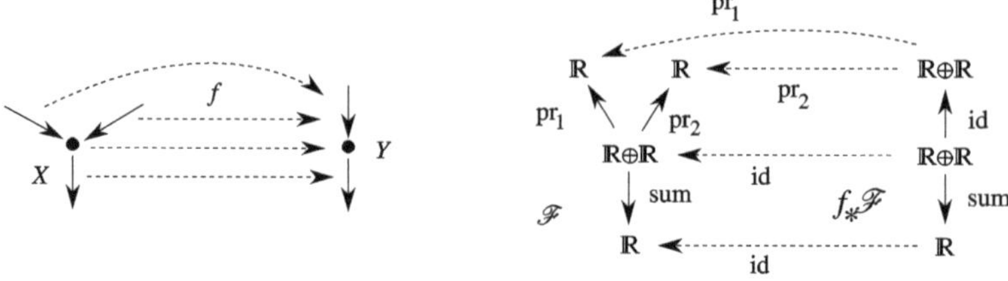

Fig. 4.7 The pushforward of sheaf $\mathscr{F}$ along a map that glues together two edges results in isomorphic sheaf cohomology

situation and implies that $\mathscr{C}$ and $f_*\mathscr{C}$ have isomorphic cohomology. This means that the information stored in the separate color channels is the same as that which is stored in the combined image. Furthermore, $f_*\mathscr{C} = \mathscr{I} \oplus \mathscr{I} \oplus \mathscr{I}$, so each *pixel* in the combined image contains information from each of the color channels.

Example 4.13 Consider the case of a flow sheaf $\mathscr{F}$ on a directed graph X as shown in Fig. 4.7. If a cellular map $f : X \to Y$ collapses the two input edges to one input edge (as shown in the figure), then the resulting pushforward sheaf $f_*\mathscr{F}$ has the same cohomology as $\mathscr{F}$ by Theorem 4.2. In this example, it is easily verified by direct computation that $H^0(X; \mathscr{F}) \cong H^0(Y; f_*\mathscr{F}) \cong \mathbb{R}^2$ and that all other cohomologies are trivial.

4.3.3 Interpretation and Examples

Recall that a global section of a sheaf $\mathscr{S}$ on X assigns a value in $\mathscr{S}(e)$ for each cell $e \in X$, provided this assignment agrees with the restrictions. Therefore, a global section is completely determined by its values on the vertices on X. Since a sheaf is also a functor, any two different compositions $\mathscr{S}(b \rightsquigarrow c) \circ \mathscr{S}(a \rightsquigarrow b)$ and $\mathscr{S}(d \rightsquigarrow c) \circ \mathscr{S}(a \rightsquigarrow d)$ must actually be the same function $\mathscr{S}(a) \to \mathscr{S}(c)$. Taken together, these two statements mean that the global sections of $\mathscr{S}$ can be recovered by restricting our attention to the 1-skeleton X^1 only. (The importance of global sections in signal processing is why most of the examples in this book are sheaves over graphs.)

Theorem 4.3 *The space of global sections of a sheaf $\mathscr{S}$ on a cell complex X is isomorphic to $H^0(X; \mathscr{S})$.*

Proof Observe that $C^0(X; \mathscr{S})$ consists of assignments of stalk values to vertices. Therefore, $\mathscr{S}(X)$ should be thought of as a subset of $C^0(X; \mathscr{S})$. It then remains to show that $\mathscr{S}(X) = \ker d^0$.

Suppose without loss of generality that X is a 1-dimensional cell complex, represented as an abstract simplicial complex with the (possible) use of an external vertex

$*$. Because we have excluded cells without compact closure, we can abuse notation slightly and suppose that $e = (v_1, v_2)$ is an edge in X. Then $[v_1 : e] = -1$ and $[v_2 : e] = 1$. If f is a global section, this means that

$$\begin{aligned}(d^k f)(e) &= \sum_{a \in X^0} [a : e]\mathscr{S}(a \rightsquigarrow e) f(a) \\ &= [v_1 : e]\mathscr{S}(v_1 \rightsquigarrow e) f(v_1) + [v_2 : e]\mathscr{S}(v_2 \rightsquigarrow e) f(v_2) \\ &= -f(e) + f(e) = 0.\end{aligned}$$

On the other hand, if $(d^k f)(e) = 0$, the same calculation implies that $\mathscr{S}(v_1 \rightsquigarrow e) f(v_1) = \mathscr{S}(v_2 \rightsquigarrow e) f(v_2)$, so the section f can be extended to e. □

The constant sheaf $\mathscr{Z}$, which assigns the group $\mathbb{Z}$ to each cell and identity maps to each attachment, is a useful tool for comparing cell complexes. For instance, the rank of the first cohomology $H^1(X; \mathscr{Z})$ can discriminate between a line segment, a circle, and several circles attached to one another at a single vertex.

Example 4.14 Consider the constant sheaf over the cell complex model of a circle with two vertices and two edges:

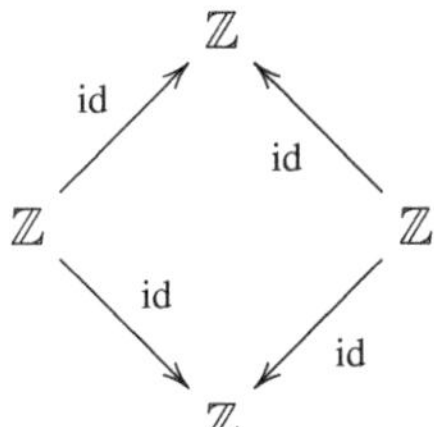

In this case, the cochain complex $(C^\bullet(X; \mathscr{Z}), d^\bullet)$ has two nonzero terms, and is given by

$$\mathbb{Z}^2 \xrightarrow{d^0 = \begin{pmatrix} 1 & -1 \\ -1 & 1 \end{pmatrix}} \mathbb{Z}^2 \longrightarrow 0$$

Observe that d^k is a homomorphism with a rank-1 kernel, so we conclude $H^0(X; \mathscr{Z}) \cong H^1(X; \mathscr{Z}) \cong \mathbb{Z}$. This agrees with the interpretation given in Theorem 4.3, since the space of global sections over the circle is also $\mathbb{Z}$.

Example 4.15 The Theorem is false for compactly supported cohomology, $H^0_c(X; \mathscr{S})$. Consider the cell complex structure

$$\text{———}\bullet\text{———}$$

on the open interval $(-1, 1)$ and the constant sheaf $\mathscr{Z}$ over it. This sheaf has the diagram

$$\mathbb{Z} \xleftarrow{\text{id}} \mathbb{Z} \xrightarrow{\text{id}} \mathbb{Z}$$

so its space of global sections is isomorphic to $\mathbb{Z}$. Consider $C_c^\bullet(X; \mathscr{S})$, which is

$$C_c^0((-1,1); \mathscr{Z}) = \mathbb{Z} \xrightarrow{d^0 = \begin{pmatrix}1\\1\end{pmatrix}} C_c^0((-1,1); \mathscr{Z}) = \mathbb{Z}^2 \longrightarrow 0.$$

The kernel of d^0 is trivial, so $H_c^0(X; \mathscr{S}) = 0$, which is not the space of global sections.

Exercise 4.5 Compute the sheaf cohomology over the circle again, but with a cell complex structure with more than 2 vertices and edges. You should find the same answer as Example 4.14.

Remark 4.3 One should be very careful to use cell complexes and not just CW complexes when manipulating sheaves. For instance, consider the CW complex structure of the circle which consists of a single vertex and edge.

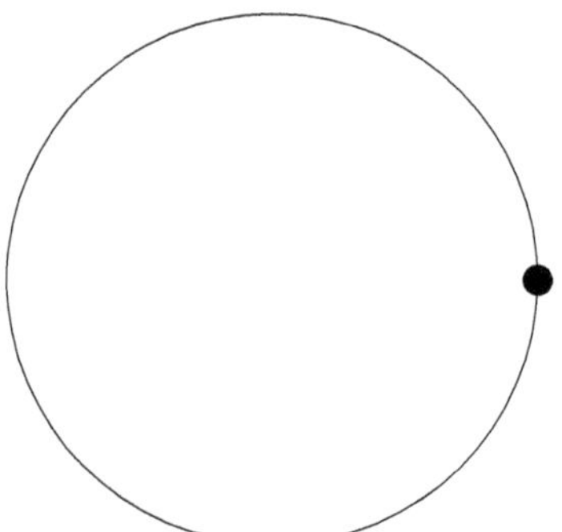

The resulting cochains fit together according to the diagram

$$\mathbb{Z} \xrightarrow{\text{id}} \mathbb{Z} \longrightarrow 0$$

which whose homology is *not* the sheaf cohomology.

Exercise 4.6 Suppose that X is a cell complex representing a collection of N circles attached to a common point. Compute the isomorphism types of $H^0(X; \mathscr{Z})$ and $H^1(X; \mathscr{Z})$.

It should be noted that the cohomology of a constant sheaf is not a perfect invariant for homotopy type of cell complexes.

Example 4.16 The real projective plane $\mathbb{RP}^n$ consists of the space of all lines through the origin in $\mathbb{R}^{n+1}$. One can easily construct $\mathbb{RP}^n$ as a CW complex by taking the quotient of the closed unit ball in $\mathbb{R}^{n+1}$ by the antipodal map which takes $x \mapsto -x$. The constant sheaves over the product spaces $\mathbb{RP}^2 \times S^3$ and $\mathbb{RP}^3 \times S^2$ have the same cohomology, but the two spaces are not homotopy equivalent. The proof that their cohomologies agree is straightforward, and is left as an exercise. To show that they are not homotopy equivalent, one can prove that their *cohomology rings* are different, which is beyond the scope of this book (the interested reader is referred to Hatcher (2002, Chapter 4)).

4.4 Long Exact Sequences for Cohomology

It is often easy to relate several sheaves to one another via a short exact sequence. The snake lemma (Lemma 4.1) creates long exact sequences from short ones. This makes it easier to reason about the cohomology of these sheaves. For instance, long exact sequences obtained in this way permit the assembly of local data over two small spaces into data over their union, and the reconstruction of a signal from its samples (reproducing the Shannon-Nyquist theorem due to Nyquist (1928), Shannon (1949) as a result).

4.4.1 Mayer–Vietoris Sequences for Sheaves

Suppose a cell complex X is formed from the union of two subcomplexes A and B, and that there is a sheaf $\mathscr{S}$ over X. We can restrict our attention to the sheaf over A and B separately; which sections on these subcomplexes extend to all of X? The pullback of $\mathscr{A}$ along the inclusion $i_A : A \to X$ is a sheaf $i_A^*\mathscr{S}$, which agrees with $\mathscr{S}$ on A. Similarly, $i_B^*\mathscr{S}$ restricts $\mathscr{S}$ to B. If we define $j_A : A \cap B \to A$ and $j_B : A \cap B \to B$ to be inclusions, then $j_A^* i_A^* \mathscr{S} = j_B^* i_B^* \mathscr{S} = i_{A\cap B}^* \mathscr{S}$.

These pieces of information can be assembled into a short exact sequence between the following sheaves

$$0 \longrightarrow \mathscr{S} \xrightarrow{i_A^* \oplus i_B^*} i_A^*\mathscr{S} \oplus i_B^*\mathscr{S} \xrightarrow{\Delta} i_{A\cap B}^*\mathscr{S} \longrightarrow 0$$

where $\Delta = j_A^* - j_B^*$. This induces the *Mayer-Vietoris long exact sequence*

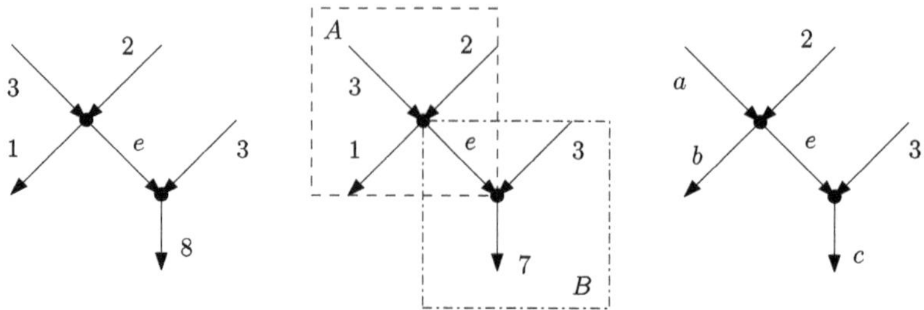

Fig. 4.8 Three sections of a flow sheaf on a graph. *Left* section cannot be extended to e; *middle* exactly one extension possible to e; *right* many extensions to e possible

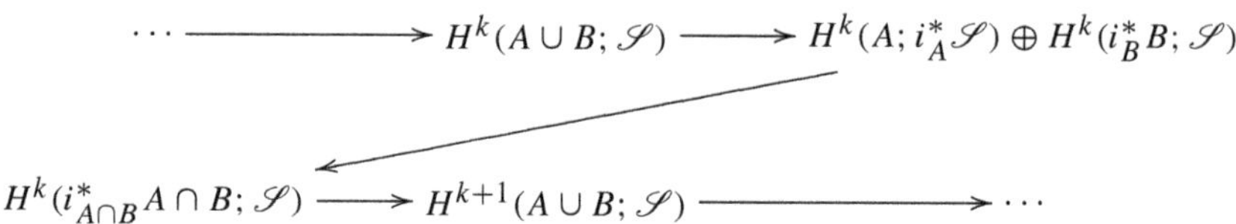

We can interpret this as stating that sections over A and B separately can extend to global sections precisely when they agree on $A \cap B$. Usually, we think of A and B initially in isolation, before later taking their union, as the following example shows.

Example 4.17 Consider the case of a flow sheaf (Example 3.6) on the directed graph shown in Fig. 4.8, with the subcomplexes A and B as marked. $A \cap B$ consists of the single edge e, where consistency between sections over A and B must be maintained for a section over A or B to extend to the whole graph. For instance, on the middle graph, the section over A (an element of $H^0(A; i_A^* \mathscr{S})$) has values of edges 3, 2, $e = 4$, 1 reading clockwise from top left. The corresponding section over B has values $e = 4$, 3, 7. The map Δ computes the difference between the sections over A and B when restricted to e. Clearly this is zero in the the middle case. Therefore, since these two sections are in the kernel of Δ, they must appear in the image of $i_A^* \oplus i_B^* : H^0(X; \mathscr{S}) \rightarrow H^0(A; i_A^* \mathscr{S}) \oplus H^0(B; i_B^* \mathscr{S})$ by exactness. Therefore they come from a single section over X.

For the case in the left frame of Fig. 4.8, observe that the section over A has the value 4 over e, but that the section over B has the value 5 there. In this case, the map Δ takes this pair of sections to 1, which is nonzero. Therefore, by exactness, this pair of sections does not come from a single section.

Finally, for the case in the right frame of Fig. 4.8, there are many possible values for e for sections over A and B separately. Several of them will result in the same value over e, and therefore lead to several possible global sections.

4.5 General Sampling Theorem for Signal Sheaves

Another important long exact sequence arises from a sheaf $\mathscr{S}$ over a cell complex X and a subcomplex Y. This exact sequence has implications for the reconstruction of a section from its values; it therefore captures the essential features of sampling. A version of the Shannon-Nyquist theorem can be proven using this long exact sequence.

Suppose that $\mathscr{M}$ is a sheaf over X that is supported on the subcomplex Y and that $m : \mathscr{S} \to \mathscr{M}$ is a sheaf morphism along the inclusion $Y \to X$. Recognize that this defines a collection of maps $m_y : \mathscr{S}(f(y)) \to \mathscr{M}(y)$. We therefore define a sheaf $\mathscr{A}$ over X whose stalks are the kernels of these maps, and whose restrictions are the restrictions of $\mathscr{S}$ restricted to the stalks of $\mathscr{A}$. We have the short exact sequence of sheaves

$$0 \longrightarrow \mathscr{A} \longrightarrow \mathscr{S} \xrightarrow{m} \mathscr{M} \longrightarrow 0$$

and by the snake lemma, we have the *long exact sequence of the pair* (X, Y)

$$\cdots \longrightarrow H^k(X; \mathscr{A}) \longrightarrow H^k(X; \mathscr{S}) \xrightarrow{m_\bullet} H^k(Y; \mathscr{M}) \longrightarrow H^{k+1}(X; \mathscr{A}) \longrightarrow \cdots$$

This sequence is particularly helpful for understanding how local sections of a sheaf extend to global ones. If we suppose that $Y \to X$ is the inclusion and $\mathscr{M}(y) = \mathscr{S}(y)$ for each cell $y \in Y$, then the global sections of $\mathscr{M}$ (elements of $H^0(Y; \mathscr{M})$) are local sections of $\mathscr{S}$. If each m_y is surjective, then $\mathscr{A}$ is isomorphic to the quotient sheaf $\mathscr{S}/\mathscr{M}$ (see Definition 3.9).

The long exact sequence indicates that if $H^0(X; \mathscr{A}) = 0$, then the kernel of $m_\bullet$ is 0. In this case, the induced map takes each global section of $\mathscr{S}$ to a unique global section of $\mathscr{M}$. On the other hand, if $H^0(X; \mathscr{A}) \neq 0$ then some global sections of $\mathscr{S}$ map to the same global section of $\mathscr{M}$. Therefore, a necessary condition to reconstruct a local signal in $\mathscr{S}$ from its samples in $\mathscr{M}$ is that $H^0(X; \mathscr{A}) = 0$. We therefore call $\mathscr{A}$ the *ambiguity sheaf* associated to the sampling morphism m.

On the other side of the exact sequence, if $H^1(X; \mathscr{A}) = 0$, then every possible set of samples can arise from a section of $\mathscr{S}$. Therefore, we have proven the following.

Theorem 4.4 *(Local signal sampling) Suppose that $m : \mathscr{S} \to \mathscr{M}$ is a sheaf morphism over an inclusion and $\mathscr{A}$ is the associated ambiguity sheaf. Global sections of $\mathscr{S}$ can be uniquely recovered from sections of $\mathscr{M}$ when $H^0(X; \mathscr{A}) = 0$. If additionally $H^1(X; \mathscr{A}) = 0$, then every section of $\mathscr{M}$ corresponds to a section of $\mathscr{S}$.*

Theorem 4.4 gives a general condition for recovery of a signal from samples. We now examine a sufficient condition (oversampling), and then a necessary condition (removal of ambiguity) for this recovery to succeed. Both of these general conditions rely on some sheaves derived from the sheaf of signals.

Definition 4.13 Suppose that $\mathscr{S}$ is a sheaf on a cell complex X. For a closed subcomplex Y of X, let $\mathscr{S}^Y$ be the sheaf whose stalks are the stalks of $\mathscr{S}$ on Y and zero elsewhere, and whose restrictions are either those of $\mathscr{S}$ on Y or zero as appropriate. There is a surjective sheaf morphism $\mathscr{S} \to \mathscr{S}^Y$ and an induced ambiguity sheaf $\mathscr{S}_Y$ which can be constructed in exactly the same way as $\mathscr{A}$ before.

Thus, the dimension of each stalk of $\mathscr{S}^Y$ is at least as large as that of any sampling sheaf, and the dimension of stalks of $\mathscr{S}_Y$ are therefore as small as or smaller than that of any ambiguity sheaf.

Proposition 4.2 *(Oversampling theorem) If X^k is the subcomplex generated by the k-cells of a cell complex X who have compact closure, then $H^k(X^{k+1}; \mathscr{S}_{X^k}) = 0$.*

Proof By direct computation, the k-cochains of $\mathscr{S}_{X^k}$ are

$$\begin{aligned} C^k(X^{k+1}; \mathscr{S}_{X^k}) &= C^k(X^{k+1}; \mathscr{S}) / C^k(X^k; \mathscr{S}) \\ &= \bigoplus_{a \text{ a } k\text{-face of } X \text{ with compact closure}} \mathscr{S}(a) \Big/ \bigoplus_{a \text{ a } k\text{-face of } X \text{ with compact closure}} \mathscr{S}(a) \\ &= 0. \end{aligned}$$

□

Corollary 4.1 *$H^0(X; \mathscr{S}_Y) = 0$ when Y is the set of vertices of X.*

Proposition 4.3 *(Sampling obstruction theorem) Suppose that Y is a closed subcomplex of a cell complex X and $m : \mathscr{S} \to \mathscr{M}$ is a sampling of sheaves on X supported on Y. If $H^0(X, \mathscr{S}_Y) \neq 0$, then the induced map $H^0(X; \mathscr{S}) \to H^0(X; \mathscr{M})$ is not injective.*

Succinctly, $H^0(X, \mathscr{S}_Y)$ is an obstruction to the recovery of global sections of $\mathscr{S}$ from its samples taken on the set Y.

Proof We begin by constructing the ambiguity sheaf $\mathscr{A}$ as before so that

$$0 \longrightarrow \mathscr{A} \longrightarrow \mathscr{S} \xrightarrow{m} \mathscr{M} \longrightarrow 0$$

is a short exact sequence. Observe that $\mathscr{M} \to \mathscr{S}^Y$ can be chosen to be injective, because the stalks of $\mathscr{M}$ have dimension not more than the dimension of $\mathscr{S}$ (and hence $\mathscr{S}^Y$ also). Thus the induced map $H^0(X; \mathscr{M}) \to H^0(X; \mathscr{S}^Y)$ is also injective. Therefore, by a diagram chase on

$$\begin{array}{ccccccc} 0 & \longrightarrow & H^0(X; \mathscr{A}) & \longrightarrow & H^0(X; \mathscr{S}) & \longrightarrow & H^0(X; \mathscr{M}) \\ & & & & \downarrow \cong & & \downarrow \\ 0 & \longrightarrow & H^0(X; \mathscr{S}_Y) & \longrightarrow & H^0(X; \mathscr{S}) & \longrightarrow & H^0(X; \mathscr{S}^Y) \end{array}$$

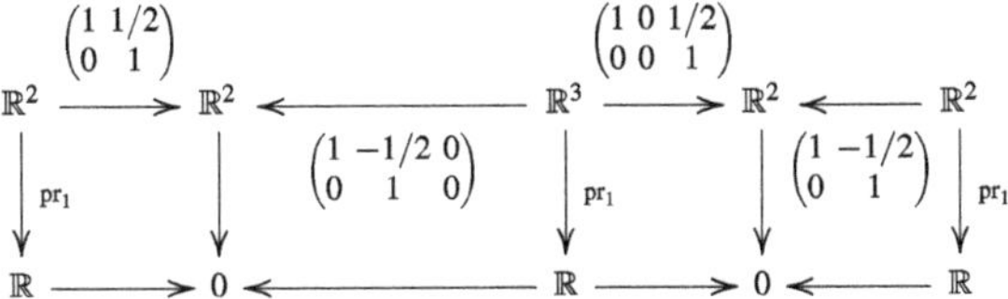

Fig. 4.9 A sheaf $\mathscr{PL}$ and a sampling of its values at the vertices

we infer that there is a surjection $H^0(X; \mathscr{A}) \to H^0(X; \mathscr{S}_Y)$. By hypothesis, this means that $H^0(X; \mathscr{A}) \neq 0$, so in particular $H^0(X; \mathscr{S}) \to H^0(X; \mathscr{M})$ cannot be injective. □

Example 4.18 Consider the sheaf $\mathscr{PL}$ (defined in Sect. 3.5) on a simplicial complex representation for the interval $[-1, 1]$, as given by the top row of the diagram in Fig. 4.9. Recall that the sections of this sheaf are continuous, piecewise linear functions whose slope may have a jump at 0. The diagram in Fig. 4.9 represents the operation of sampling the values of these functions at the three vertices. In this case, the ambiguity sheaf $\mathscr{A}$ is

$$0 \longrightarrow \mathbb{R}^2 \xleftarrow{\begin{pmatrix}-1/2 & 0\\ 1 & 0\end{pmatrix}} \mathbb{R}^2 \xrightarrow{\begin{pmatrix}0 & 1/2\\ 0 & 1\end{pmatrix}} \mathbb{R}^2 \longleftarrow 0$$

Intuitively, sampling a continuous, piecewise linear function at each of its vertices should allow for perfect reconstruction of the function. To verify that this agrees with Proposition 4.3, we therefore compute the coboundary map d^0 associated to $\mathscr{A}$, which is

$$d^0 = \begin{pmatrix} 0 & 1/2 \\ 0 & 1 \\ 1/2 & 0 \\ -1 & 0 \end{pmatrix}$$

The kernel of this matrix is trivial so $H^0(\mathscr{A}) = 0$, which implies that sampling is unambiguous.

4.5.1 The Shannon–Nyquist Theorem

As a special case of the general sampling theorems, we prove the traditional Shannon-Nyquist theorem, by considering a sheaf of bandlimited signals on the real line $\mathbb{R}$.

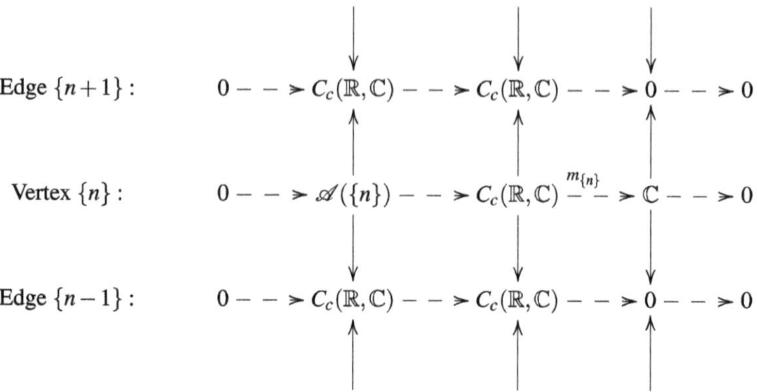

Fig. 4.10 The short exact sequence of sheaves for the Nyquist theorem, in which the *dotted arrows* indicate sheaf morphisms. The *middle* column represents the sheaf of bandlimited functions, the *right* sheaf represents its samples, and the *left* sheaf represents the resulting ambiguity sheaf

We will use the cell complex structure X given by $X^0 = \mathbb{Z}$ and $X^1 = \{(n, n+1)\}$. We construct the sheaf $\mathscr{C}$ of bandlimited signals in a short exact sequence, as shown in Fig. 4.10 and detailed below.

- For every cell, we represent the bandlimited signals on that cell by their Fourier transform. Therefore, let every stalk of $\mathscr{C}$ be $C_c(\mathbb{R}, \mathbb{C})$, the set of compactly supported complex-valued continuous functions.
- Let each inclusion $\mathscr{C}(\{n\} \rightsquigarrow \{n, n+1\})$ and $\mathscr{C}(\{n+1\} \rightsquigarrow \{n, n+1\})$ be the identity.

Then $H^0(\mathbb{R}; \mathscr{C})$ is just $C_c(\mathbb{R}, \mathbb{C})$. Construct the sheaf $\mathscr{M}$ whose stalk on each vertex is $\mathbb{C}$ and each edge stalk is zero. We construct a morphism by the zero map on each edge, and by the inverse Fourier transform as shown below on vertex $\{n\}$

$$m_{\{n\}}(f) = \int_{-\infty}^{\infty} f(\omega)e^{-2\pi i n\omega} d\omega.$$

Then the ambiguity sheaf $\mathscr{A}$ has stalks $C_c(\mathbb{R}, \mathbb{C})$ on each edge, and

$$\mathscr{A}(\{n\}) = \{f \in C_c(\mathbb{R}, \mathbb{C}) : \int_{-\infty}^{\infty} f(\omega)e^{-2\pi i n\omega} d\omega = 0\}$$

on each vertex $\{n\}$.

The sections of $H^0(X; \mathscr{A})$ consist of global sections of $\mathscr{C}$ for which

$$\int_{-\infty}^{\infty} f(\omega)e^{-2\pi i n\omega} d\omega = 0 \text{ for all } n.$$

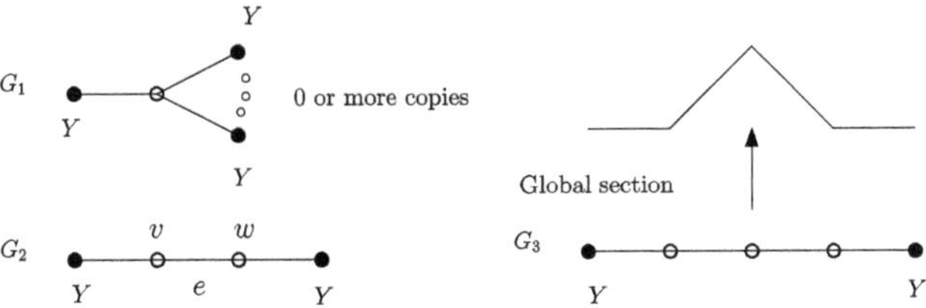

Fig. 4.11 Graphs G_1, and G_2 (*left*) and G_3 (*right*) for Lemma 4.3. Filled vertices represent elements of Y, empty ones are in the complement of Y

Because many functions satisfy this collection of equations, the obstruction to reconstruction is fairly large. However, the Shannon-Nyquist theorem argues that if a global band limit is chosen, the obstruction vanishes.

Theorem 4.5 *(Traditional Shannon-Nyquist theorem) Suppose we replace $C_c(\mathbb{R}, \mathbb{C})$ with the set of continuous functions supported on $[-B, B]$. Then if $B \leq 1/2$, the resulting ambiguity sheaf $\mathscr{A}$ has $H^0(X; \mathscr{A}) = 0$.*

Proof The elements of $H^0(X; \mathscr{A})$ are now given by the compactly supported continuous functions f on $[-B, B]$ for which

$$\int_{-B}^{B} f(\omega)e^{-2\pi i n \omega} d\omega = 0 \text{ for all } n.$$

Observe that if $B \leq 1/2$, this is precisely the statement that the Fourier series coefficients of f all vanish; hence f must vanish. This means that the only global section of $\mathscr{A}$ is the zero function. (Ambiguities can arise if $B > 1/2$, because the set of functions $\{e^{-2\pi i n \omega}\}_{n\in\mathbb{Z}}$ is then *not* complete.) □

4.5.2 Sampling of Heterogeneous, Non-bandlimited Signals

The real power of using sheaves in sampling theory is their generality. Besides band limited signals, there are other classes of signals that can be reconstructed from their samples. The sheaf-theoretic Nyquist theorem can treat nontrivial base space topologies as well as samples of different dimensions. Consider the example of the sheaf of piecewise linear functions $\mathscr{PL}$ on a graph introduced in Sect. 3.5 and the sampling morphism $s : \mathscr{PL} \to \mathscr{PL}^Y$ where Y is a subset of the vertices of X. Excluding one or two vertices from Y does not prevent reconstruction in this case, because the samples include information about slopes along adjacent edges.

Lemma 4.3 *Consider $\mathscr{PL}_Y$, the subsheaf of $\mathscr{PL}$ whose sections vanish on a vertex set Y and the graphs G_1, G_2, and G_3 as shown in Fig. 4.11. There are no*

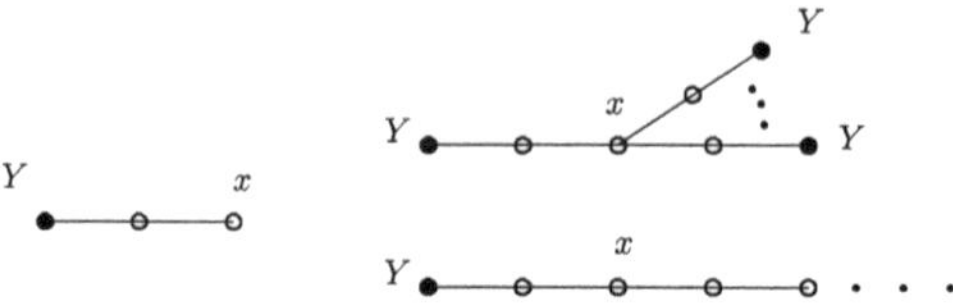

Fig. 4.12 The three families of subgraphs that arise when med $(Y) > 1$. Filled vertices represent elements of Y, empty ones are in the complement of Y

nontrivial sections of $\mathscr{PL}_Y$ *on* G_1 *and* G_2*, but there are nontrivial sections of* $\mathscr{PL}_Y$ *on* G_3.

Proof If a section of $\mathscr{PL}$ vanishes at a vertex x with degree n, this means that the value of the section there is an $(n+1)$-dimensional zero vector. The value of the section on every edge adjacent to x is then the 2-dimensional zero vector. Since the dimensions in each stalk of $\mathscr{PL}$ represent the value of the piecewise linear function and its slopes, linear extrapolation to the center vertex in G_1 implies that its value is zero too.

A similar idea applies in the case of G_2. The stalk at v has dimension 3. Any section at v that extends to the left must actually lie in the subspace spanned by $(0, 0, 1)$ (coordinates represent the value, left slope, right slope respectively). In the same way, any section at w that extends to the right must lie in the subspace spanned by $(0, 1, 0)$. Any global section must extend to e, which must therefore have zero slope and zero value.

Finally G_3 has nontrivial global sections, spanned by the one shown in Fig. 4.11. $\square$

Definition 4.14 On a graph G, define the *edge distance* between two vertices v, w to be

$$\mathrm{ed}(v, w) = \begin{cases} \min_p \{\# \text{ edges in } p \text{ where } p \text{ is a PL-continuous path from } v \to w\} \\ \infty \ \text{ if no such path exists} \end{cases}$$

From this, the maximal distance to a vertex set Y is

$$\mathrm{med}\,(Y) = \max_{x \in X^0} \{\min_{y \in Y} \mathrm{ed}(x, y)\}.$$

Proposition 4.4 *(Unambiguous sampling) Consider the sheaf* $\mathscr{PL}$ *on a graph* X *and* $Y \subseteq X^0$. *Then* $H^0(X; \mathscr{F}_Y) = 0$ *if and only if* $\mathrm{med}\,(Y) \leq 1$.

Proof ($\Leftarrow$) Suppose that $x \in X^0 \backslash Y$ is a vertex not in Y. Then there exists a path with one edge connecting it to Y. Whence we are in the case of G_1 of Lemma 4.3, so any section at x must vanish.

($\Rightarrow$) By contradiction. Assume med $(Y) > 1$. Without loss of generality, consider $x \in X^0 \backslash Y$, whose distance to Y is exactly 2. Then one of the subgraphs shown in

Fig. 4.12 must be present in X. But case G_3 of Lemma 4.3 makes it clear that the most constrained of these (the middle panel of Fig. 4.12) has nontrivial sections at x, merely looking at sections over the subgraph. □

Proposition 4.5 *(Non-redundant sampling of $\mathscr{PL}$) Consider the case of $s : \mathscr{PL} \to \mathscr{PL}^Y$. If $Y = X^0$, then $H^1(X; \mathscr{A}) \neq 0$. If Y is such that med $(Y) \leq 1$ and $|X^0 \backslash Y| + \sum_{y \notin Y} \deg y = 2|X^1|$, then $H^1(X; \mathscr{A}) = 0$.*

Proof The stalk of $\mathscr{A}$ over each edge is $\mathbb{R}^2$, and the stalk over a vertex in Y is trival. However, the stalk over a vertex of degree n not in Y is $\mathbb{R}^{n+1}$. Observe that if $H^0(X; \mathscr{A}) = 0$, then $H^1(X; \mathscr{A}) = C^1(X; \mathscr{A})/C^0(X; \mathscr{A})$. Using the degree sum formula in graph theory, we compute that $H^1(X; \mathscr{A})$ has dimension $2|X^1| - \sum_{y \notin Y}(\deg y + 1)$. □

4.5.3 Sampling in Topological Filters

We now tie up a loose end from Sect. 3.4.1, that of the proper definition of a topological filter.

Definition 4.15 A diagram of sheaf morphisms

$$\mathscr{S}_1 \xleftarrow{m_1} \mathscr{S}_2 \xrightarrow{m_2} \mathscr{S}_3$$

is called a *topological filter* from $\mathscr{S}_1 \to \mathscr{S}_3$ when m_1 induces isomorphisms on sheaf cohomology. We call $\mathscr{S}_1$ the *input*, $\mathscr{S}_3$ the *output*, and $\mathscr{S}_2$ the *state* of the topological filter.

This definition generalizes the properties already seen in Sect. 3.4.1 and proven in Proposition 3.4. Specifically, *all linear translation-invariant filters are topological filters*. However, the class of topological filters is strictly larger. One need only consider an appropriate diagram of sheaves over a graph in order to obtain a filter for which "translation invariance" has no meaning.

Example 4.19 An example of a topological filter that is not also a translation invariant system is shown in Fig. 4.13. The sheaf on the left and the sheaf in the middle have isomorphic cohomology, and the morphism between them is an isomorphism. Therefore, according to Definition 4.15, the sheaf on the left represents the filter's input signal. The middle sheaf (representing the internal state of the filter) plays the role of reformatting the input in a way that is convenient to translate it to the output sheaf on the right. The filter removes information, since the dimension of the space of global sections of $\mathscr{F}_1$ is larger than that of $\mathscr{F}_2$.

Remark 4.4 Readers already familiar with sheaf theory will recognize a topological filter as being a special kind of morphism in the derived category of sheaves (see Shepard 1980, Chapter 2). The usual definition also applies to chain complexes

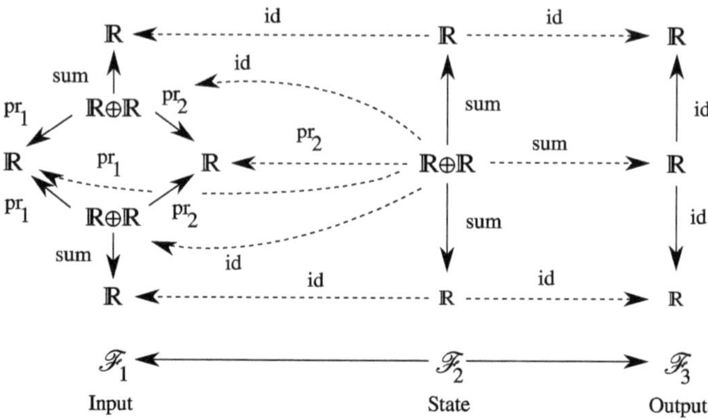

Fig. 4.13 A topological filter which is not translation invariant

(Definition 4.6) of sheaves and chain maps as well, though we will not require this level of generality.

The ambiguity sheaf can be extended to treat topological filters. Suppose $f : \mathscr{S} \to \mathscr{M}$ is a topological filter. This means that there are two sheaf morphisms

$$\mathscr{S} \xleftarrow[f_1]{} \mathscr{R} \xrightarrow{f_2} \mathscr{M}$$

in which f_1 induces an isomorphism on H^0. Let us suppose that $\mathscr{S}$, $\mathscr{R}$, and $\mathscr{M}$ are sheaves on X, Y, and Z, respectively. Ambiguities in sampling arise when the kernel K of the map $f_{2\bullet} : H^0(Y; \mathscr{R}) \to H^0(Z; \mathscr{M})$ induced by f_2 is nontrivial. Since f_1 induces an isomorphism $H^0(Y; \mathscr{R}) \to H^0(X; \mathscr{S})$, this means that $f_{2\bullet}(K)$ is a nontrivial subspace of the space of global sections of $\mathscr{S}$.

We can recover the ambiguity sheaf using the following procedure. Let $\mathscr{A}'$ be the ambiguity sheaf constructed so that

$$0 \longrightarrow \mathscr{A}' \xrightarrow{i} \mathscr{R} \longrightarrow \mathscr{M}$$

is exact, with i being a sheaf morphism over the identity map. Therefore, $\mathscr{A}'$ is a subsheaf of $\mathscr{R}$, and so we can obtain $\mathscr{A}$ as the image of $\mathscr{A}'$ through f_1, a subsheaf of $\mathscr{S}$. By construction, the space of global sections of $\mathscr{A}$ is isomorphic to $f_{2\bullet}(K)$. In other words, we have the following necessary condition for recovery of samples from a topological filter.

Proposition 4.6 *The input to a topological filter can be recovered from its output when the global sections of its ambiguity sheaf* $\mathscr{A}$ *are all zero.*

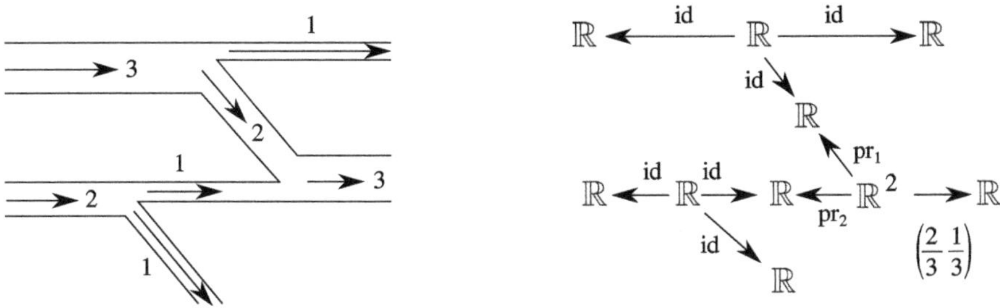

Fig. 4.14 A water flow network (*left*), and the concentration sheaf on this network (*right*)

4.6 Case Study: Tracking Water Pollution

The abundance of clean drinking water is a hallmark of civilization. Bringing clean water into a city away from sources of pollution has presented an important engineering challenge from the earliest times. Contamination of water sources is a serious public health problem, for which penalties have been established. However, once a contamination has occurred, it can be difficult to establish blame conclusively. The sampling theory for sheaves provides a useful framework for ensuring that contaminant concentration levels are sufficiently measured and that polluters are located without any doubt.

4.6.1 A Sheaf of Concentrations

The flow of water through a network of channels can be represented by an oriented 1-dimensional cell complex, whose edges are labelled by a real-valued function F representing the volume flow rate of water over that edge (in units such as liters per second). We assume that F satisfies a conservation law: the sum of flow rates into a vertex equals the sum of the flow rates out of it. See Fig. 4.14 for an example of cell complex labeled with flow rates.

Remark 4.5 The possible flow rate functions on a cell complex X are given by the global sections of the flow sheaf on X, as described in Example 3.6.

The flow of a contaminant carried by moving water can be described by a collection of measurements of its concentration. We assume that the contaminant mixes thoroughly at each junction. From a modeling perspective, this means that the mass of the contaminant is conserved at each vertex in X and that the concentration along each edge exiting a given vertex is the same. This leads to the following definition of a sheaf that records the concentration of a contaminated water network.

Definition 4.16 The *concentration sheaf* $\mathscr{C}$ on X with flow F has stalk

- $\mathscr{C}(v) = \mathbb{R}^n$ over each vertex v of X with indegree n, and
- $\mathscr{C}(e) = \mathbb{R}$ over each edge e of X.

Suppose that v is a vertex in X, that $\{e_1^+, \ldots, e_n^+\}$ are the incoming edges attached to v, and that $\{e_1^-, \ldots, e_m^-\}$ are the outgoing edges attached to v. Suppose that $(c_1, \ldots, c_n) \in \mathscr{C}(v)$ are the concentrations on each of the incoming edges, so that

$$\left(\mathscr{C}(v \rightsquigarrow e_i^+)\right)(c_1, \ldots, c_n) = c_i.$$

Because of the perfect mixing assumption, the concentration along each outgoing edge is the same value, namely

$$\left(\mathscr{C}(v \rightsquigarrow e_i^-)\right)(c_1, \ldots, c_n) = \frac{\sum_{j=1}^n F(e_j^+)c_j}{\sum_{j=1}^n F(e_j^+)}.$$

Example 4.20 Consider the network shown at left in Fig. 4.14. Its corresponding concentration sheaf is shown at right. What should be immediately clear is that portions of the network have the same concentration. The place where the concentration changes abruptly is at the lower right vertex. The outflow from this "mixing point" has a concentration that is a weighted average of the concentration of its inflows.

4.6.2 Elementary Water Flow Networks

Example 4.20 shows that the concentration of a contaminant carried by a water flow network changes abruptly at specific locations. This fact has implications for water quality analysis and tracking. Consider the case of a drinking water distribution network that draws water from a variety of sources and delivers it to a collection of customers. If the network is contaminated at a specific location, can this location be recovered by measurements at the point of delivery? Or is it necessary to monitor water quality internal to the network as well?

In order to address this problem, we make use of the ambiguity sheaf associated to sampling the concentration sheaf for the network. The cohomology of the ambiguity sheaf quantifies both

1. the ways that a contaminated water network can incorrectly appear to be clean and
2. the ways to assign blame for polluting the network

given a set of concentration measurements.

The concentration sheaves associated to the two networks shown in Fig. 4.15 are isomorphic. Any network can be decomposed into a union of networks like the two shown in Fig. 4.16. We analyze the sampling of water quality on these two networks to understand sampling larger networks.

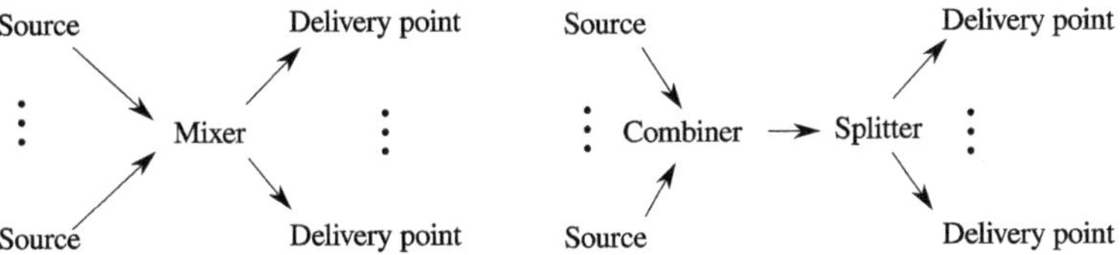

Fig. 4.15 The concentration sheaves for these two distribution networks are isomorphic

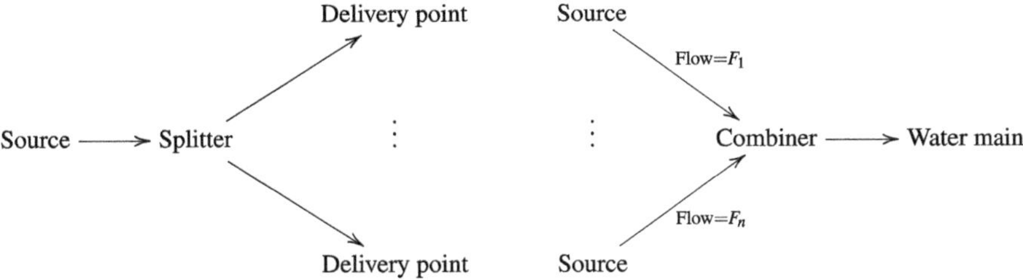

Fig. 4.16 Splitting a single water source among several points of delivery (*left*), combining several water sources (*right*)

Exercise 4.7 Construct the sheaf isomorphism between the concentration sheaves for the networks shown in Fig. 4.15.

The network on the right of Fig. 4.16 cannot be completely characterized by measurements at the single delivery point. Specifically, if no contaminant is found at the delivery point, then neither source was contaminated. However, if the delivery point is contaminated, it is impossible to tell which source was the culprit, or if both were contaminated. To make that inference, we must measure at the sources. To demonstrate how the sheaf-theoretic sampling methodology works, observe that the concentration sheaf for the network is given by the following diagram

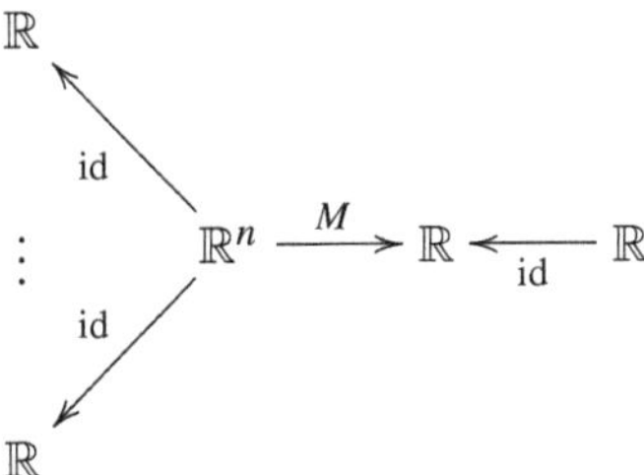

where M is given by a matrix that depends on the flow rates into the combiner. The sampling sheaf associated to measuring at the delivery point has the diagram

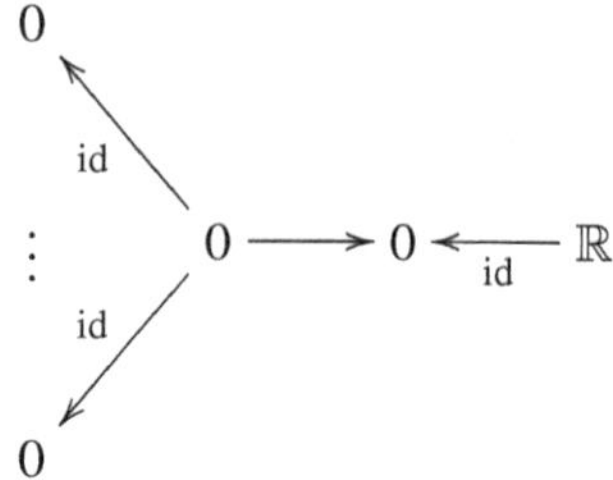

This leads to the following ambiguity sheaf $\mathscr{A}$

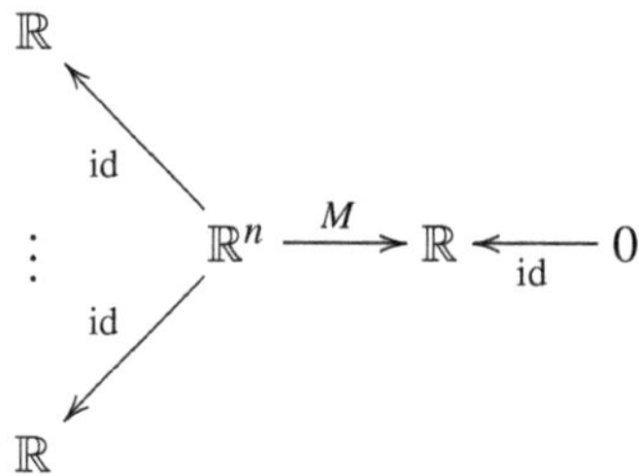

which has coboundary map

$$d^0 = \begin{pmatrix} \frac{F(e_1)}{F(e_{n+1})} & \frac{F(e_2)}{F(e_{n+1})} & \cdots & \frac{F(e_n)}{F(e_{n+1})} \end{pmatrix}.$$

This matrix has a kernel with dimension $n-1$, which is also the dimension of $H^0(\mathscr{A})$. Each element of this space of global sections can be generated by elements like

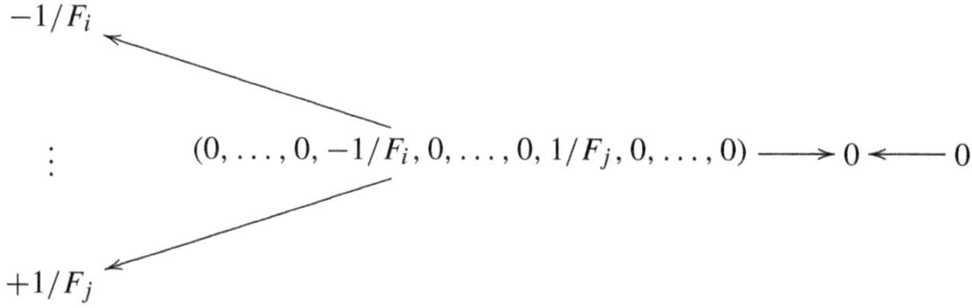

This indicates that concentration at the output can be held constant by reducing the concentration at one source and increasing it at the other. Referring back to the original concentration sheaf, if the outflow concentration is zero, then no one is polluting. However, if the outflow concentration is not zero, it is impossible to tell which source is polluting. Sadly many rivers are like this, which makes tracking contaminants and assigning blame impossible without additional measurements.

The network on the left of Fig. 4.16 can be measured completely at its delivery points. This structure is very common in drinking water distribution networks, where a single water main is the source, and each delivery point represents a water customer. In this case, the concentration sheaf has the particularly simple structure given by the diagram

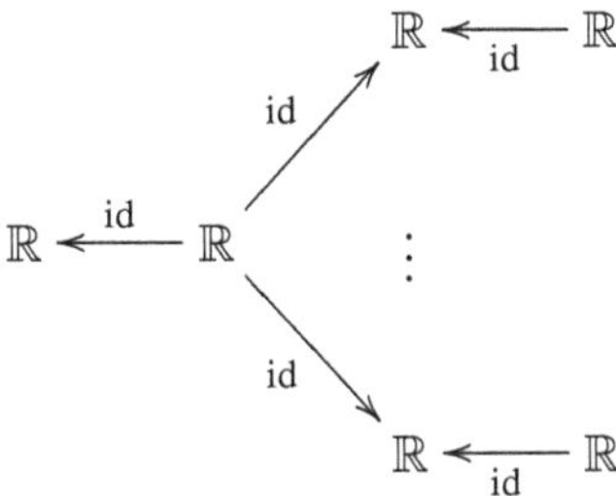

By inspection, the space of sections is determined uniquely by measurement at any one of the delivery points. Therefore, the contaminant concentration at the source can be recovered exactly from a measurement at any of the delivery points. Because of this reason, it is easy to ensure safe drinking water in a sealed distribution system (and track contaminations to their source) by endpoint checks.

Exercise 4.8 Apply the sheaf-theoretic sampling formalism to the network on the left of Fig. 4.16. Specifically,

1. Construct the sampling sheaf that represents measuring the contaminant concentration at one of the delivery points,
2. Construct the associated ambiguity sheaf $\mathscr{A}$ for your sampling sheaf,
3. Write the coboundary map $d^0 : C^0(\mathscr{A}) \to C^1(\mathscr{A})$ as a matrix, and
4. Verify that the kernel of d^0 is trivial.

Exercise 4.9 Consider the network

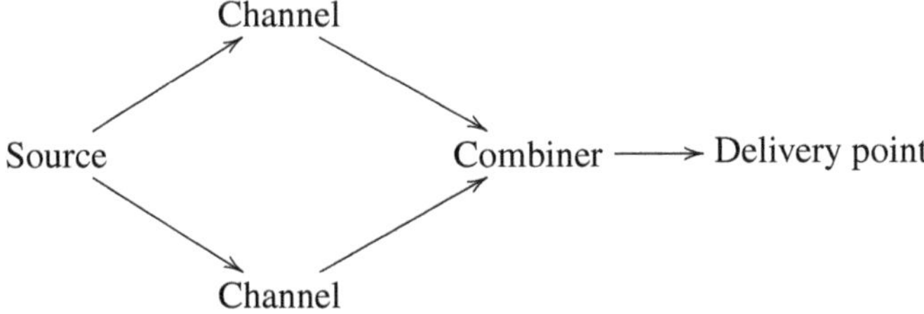

1. Construct the concentration sheaf over this network (with flow rates left as variables).
2. Compute cohomology of the ambiguity sheaf sampled at each of the Source, Channel, and Combiner.
3. Interpret your results in terms of water quality measurement.

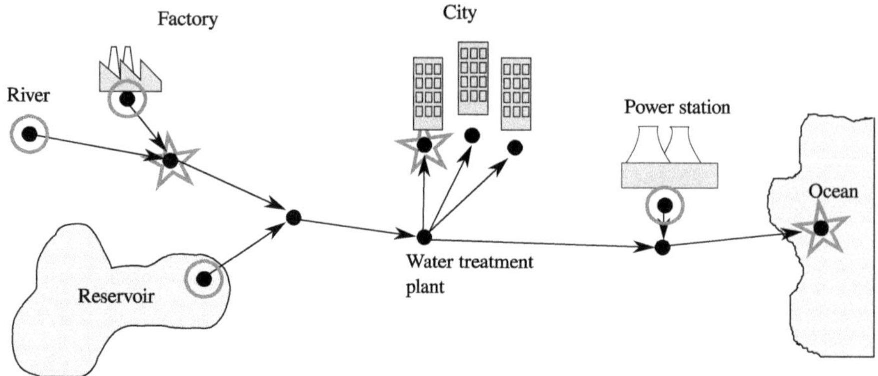

Fig. 4.17 A water distribution network including a city, reservior, and industrial areas. Two sets of water quality monitoring stations are marked: circles (water stream inputs) and stars (key nodes)

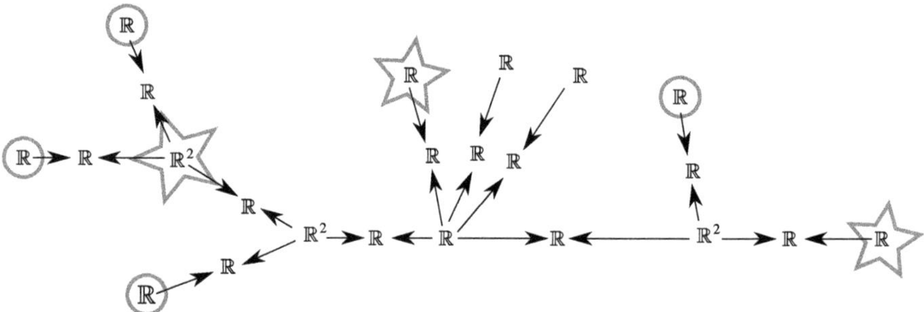

Fig. 4.18 The concentration sheaf associated with the water distribution network shown in Fig. 4.17

4.6.3 Measurement of Larger Networks

Consider the water distribution network shown in Fig. 4.17. This network contains a number of sources, delivery points, and internal junctions. We will consider the implication of sampling either at each source (points marked with circles) or at a few key locations (points marked with stars). Although sampling at the water treatment plant certainly ensures safe drinking water for the city, it provides no way to conclusively detect pollution from the factory or the power plant. If pollution is detected, it is unclear whether it is from the factory, the reservoir, or from the river. Worse, pollution from the power plant will go undetected. This model illustrates why it is difficult to ensure compliance with environmental regulations; they fundamentally require some amount of cooperation from the polluters.

Consider the case of measuring the concentration at each source (the circles in Fig. 4.17 using the concentration sheaf in Fig. 4.18). To obtain the ambiguity sheaf from Fig. 4.18, it suffices to set each circled stalk to zero. The resulting ambiguity sheaf's coboundary map has the form

$$\begin{pmatrix}
1 & & & & & & & & & \\
 & 1 & & & & & & & & \\
* & * & 1 & & & & & & & \\
 & & & 1 & & & & & & \\
 & & * & * & 1 & & & & & \\
 & & & & 1 & & & & & \\
 & & & & 1 & 1 & & & & \\
 & & & & 1 & & 1 & & & \\
 & & & & 1 & & & 1 & & \\
 & & & & & & & & 1 & \\
 & & & & & & & * & * & 1
\end{pmatrix}$$

which is of full rank. Hence there is no ambiguity that arises from sampling water quality at each source.

This is perhaps wasteful in that it requires more measurements than is necessary as the next exercise shows.

Exercise 4.10 Show, by computing the cohomology of the ambiguity sheaf, that sampling at the starred nodes in Fig. 4.17 is sufficient to completely determine the concentration at all points in the network.

4.7 Case Study: Extracting Topology from Intersections in Coverage

We continue the case studies from Chaps. 2 and 3, in which we study the topology of a wave propagating environment with multiple transmitters and receivers. By using information about the shadows cast by various obstacles for different transmitters, it is possible to assemble a cell complex model of the scene. Under favorable circumstances, this model is homotopy equivalent to the scene, where the occlusions are represented as being regions where propagation is disallowed.

4.7.1 The Nerve Model of a Space

In order to construct the model of a propagating environment for transmitters and receivers, we need to study the structure of open covers. We assign an open set U_i to each transmitter i, which represents its *coverage region*, the region on which its signal can be reliably received. We will assume that the collection $\mathscr{U} = \{U_1, \ldots, U_n\}$ of open sets forms a *cover* of the entire propagating environment X, by which we mean $\bigcup_i U_i = X$.

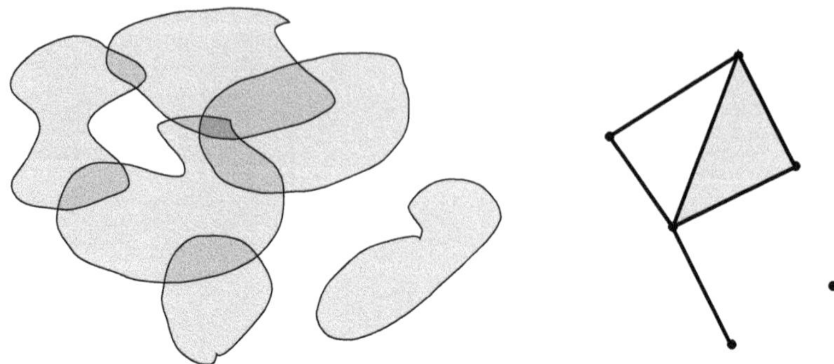

Fig. 4.19 An open cover $\mathscr{U}$ (*left*) and its nerve (*right*)

Open covers play an important, general role in the topological study of spaces. The notion of a *nerve* (Fig. 4.19) connects open covers to the homotopy type of a topological space.

Definition 4.17 Suppose that $\mathscr{U}$ is an open cover of a topological space X. The *nerve of* $\mathscr{U}$ (written $N(\mathscr{U})$) is the abstract simplicial complex on the elements of $\mathscr{U}$ (the abstract vertices) in which there is a k-simplex $\{U_1, \ldots, U_{k+1}\}$ for each nonempty intersection $U_1 \cap \cdots \cap U_{k+1}$ of $k+1$ sets $U_1, \ldots, U_{k+1} \in \mathscr{U}$.

Because $N(\mathscr{U})$ is an abstract simplicial complex (see Sect. 2.2.1), it can be realized (Definition 2.11) as a CW complex in which each attachment corresponds to a subset relation. If a simplex $\tau = \{U_{i_1}, \ldots, U_{i_n}\}$ is attached to $\sigma = \{U_{j_1}, \ldots, U_{j_m}\}$ in the nerve of a cover $\mathscr{U}$, this means that as sets of open sets

$$\{U_{i_1}, \ldots, U_{i_n}\} \subset \{U_{j_1}, \ldots, U_{j_m}\}.$$

Therefore $m > n$, so the dimension of σ is larger than that of τ. However, as subsets of X, observe that

$$U_{j_1} \cap \cdots \cap U_{j_m} \subset U_{i_1} \cap \cdots \cap U_{i_n}.$$

The intersection of open subsets of X corresponding to the higher dimensional simplex σ is a subset of the lower dimensional simplex τ.

If the transmitters can be uniquely identified by the receivers, it is straightforward to assemble a nerve from the coverage regions $\mathscr{U} = \{U_i\}$. The presence of a nonempty intersection between coverage regions is *witnessed* by the existence of a receiver detecting several transmitters simultaneously.

If a receiver r_j detects the transmitter at t_i, then $r_j \in U_i$. The nerve can then be assembled using only knowledge of the r_j and their membership in the coverage regions. Practical systems often ascertain whether a receiver detects a transmitter or not by testing whether the received signal strength from that transmitter is higher than a specified *detection threshold* D. Additionally, it is possible for noise to cause a receiver to falsely detect a transmitter. Therefore, we assume that $U_{i_1} \cap \cdots \cap U_{i_k}$ is nonempty if and only if there is at least a minimum number of r_j in that intersection. This minimum number is called the *witness threshold*. Using this, we arrive at the *nerve* $N(\mathscr{U}, R)$ *of* $\mathscr{U}$ *witnessed by* $R = \{r_j\}_{j=1}^m$, which can be a good approximation to the nerve of $\mathscr{U}$.

An algorithm for computing the nerve associated to a set of signal strength measurements can be defined as follows.

Algorithm 2. (Computing the nerve of a cover)
Input:

1. The witness threshold W
2. The detection threshold D
3. A list of transmitters T and a list of receivers R
4. An $n \times m$ matrix of signal strengths $s_{i,j}$ received by $r_j \in R$ from transmitter $t_i \in T$

Output: A list of lists of transmitters specifying the simplices the nerve $N(\mathscr{U}, R)$
Procedure:

1. $Nerve \leftarrow \{\}$
2. For each subset $\{\tau_1, \ldots, \tau_k\} \subseteq \{1, \ldots, n\}$,
   a. $counter \leftarrow 0$
   b. For $j = 1$ to m,
      i. If all of the $s_{\tau_i, j} > D$ for $i = 0$ to k, then increment $counter$
   c. If $counter > W$, then append $\{\tau_1, \ldots, \tau_k\}$ to $Nerve$
3. Return $Nerve$

This algorithm is not the most efficient one possible, because it iterates over all possible simplices formed with transmitters as vertices. This set is quite large, and can be managed by placing an upper bound on the dimension of simplices that will be constructed.

Exercise 4.11 Describe how to modify Algorithm 2 for a general open cover. Hint: Only one line really needs to be changed.

Example 4.21 Consider the floorplan of the third floor of the David Rittenhouse Laboratory of the University of Pennsylvania shown in Fig. 4.20. There are eight wireless network transmitters located on the walls as indicated in the figure. Using a simple model of radio propagation, the received signal strength from each transmitter was simulated on a dense grid of points spaced every 0.5 m. The signal strength for transmitter 1 is shown in grayscale in Fig. 4.20. This simulated the process of sampling the environment by moving a receiver to many locations.

Using these predicted measurements, the matrix of signal strengths was computed and is shown in Fig. 4.21. Note in particular that the matrix has eight rows and many columns, one for each location where the receiver was placed. The banded structure in this matrix is an artifact of how the receiver was raster-scanned through the hallways.

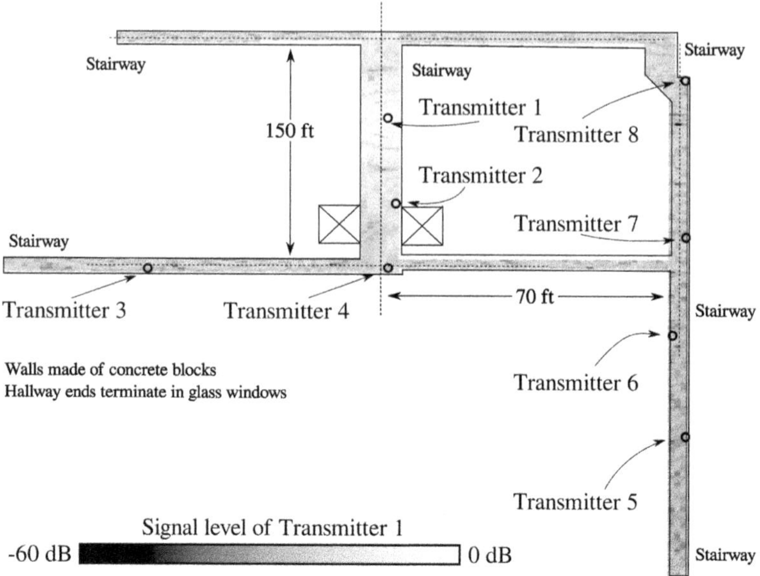

Fig. 4.20 Dimensioned floorplan of the third floor of David Rittenhouse Laboratory with the received signal strength of transmitter 1 shown

The output of Algorithm 2 for several choices of witness and detection thresholds is shown in Figs. 4.22 and 4.23. It is difficult to visualize the output nerves, which are abstract simplicial complexes. Since the locations of the transmitters were known in this simulation, they have been used to specify the locations of the vertices of the nerves. This makes the relationship between the topology of the nerve and the topology of the floorplan easier to see. In practice, this information may not be available.

For certain choices of thresholds (as in Fig. 4.22), the nerve correctly recovers the fact that there is a walled courtyard in the building, through which radio signals will not propagate.

Figure 4.23 shows small changes in the thresholds can result in substantially different nerves. This means that Algorithm 2 is not very robust; practical implementations require that this defect be mitigated. An important first step is to establish theoretical guarantees on correct performance of the Algorithm. We cannot hope for a geometric condition based on signal-to-noise ratio or the distances between measurements, since the nerve only depends on the intersections of open sets. Instead, there is a topological statement about the quality of the overall transmitter and receiver placements. Although a complete characterization of placements that yield topologically accurate nerves is elusive, there is a famous sufficient condition called the Nerve Lemma, originally due to Borsuk (1948).

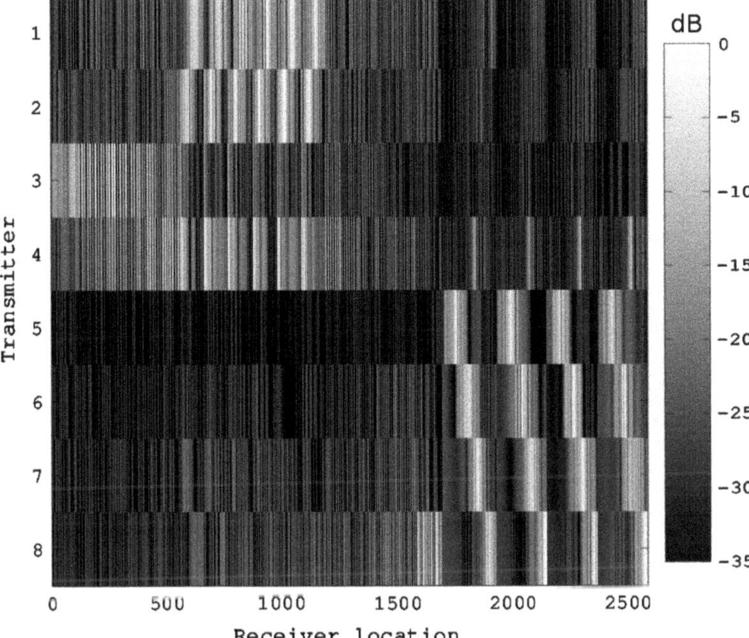

Fig. 4.21 Simulated received signal strength matrix for the third floor of David Rittenhouse Laboratory

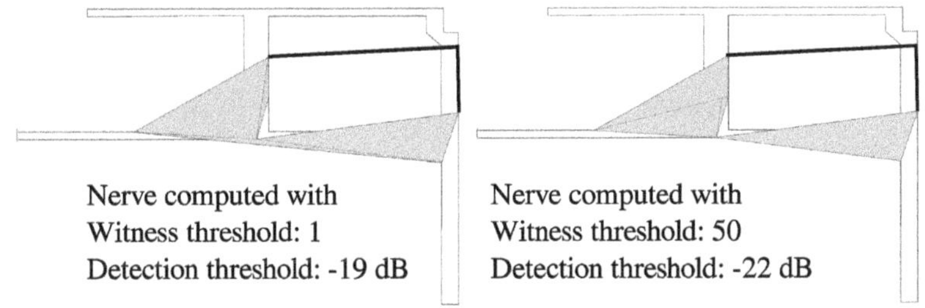

Fig. 4.22 Two outputs of Algorithm 2. *Left* witness threshold at 1 location and detection threshold at $-19\,\text{dB}$. *Right* witness threshold at 50 locations and detection threshold at $-22\,\text{dB}$

Theorem 4.6 *(the Nerve Lemma) Let $\mathbb{Z}$ be the constant sheaf over a cell complex X. If $\mathscr{U}$ is an open cover of X in which every intersection between sets in $\mathscr{U}$ is contractible, then the cohomology of the constant sheaf over the nerve $H^\bullet(N(\mathscr{U}); \mathbb{Z})$ is isomorphic to $H^\bullet(X; \mathbb{Z})$.*

The nerve lemma requires the use of *singular cohomology* and is proven in Bott and Tu (1995, Thm. 8.9 and Prop. 10.6).

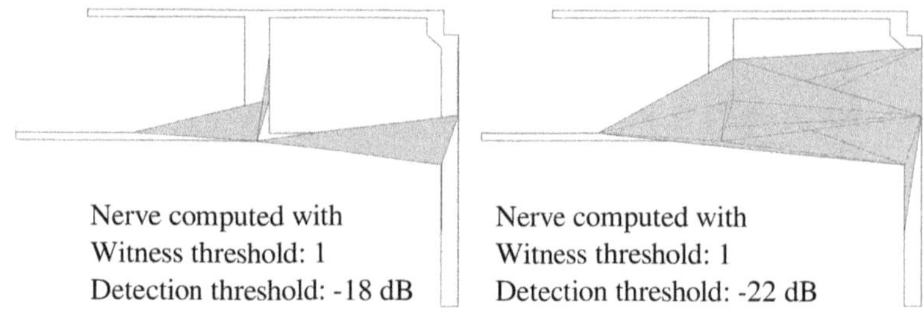

Fig. 4.23 Two outputs of Algorithm 2. *Left* witness threshold at 1 location and detection threshold at -18 dB. *Right* witness threshold at 1 location and detection threshold at -22 dB

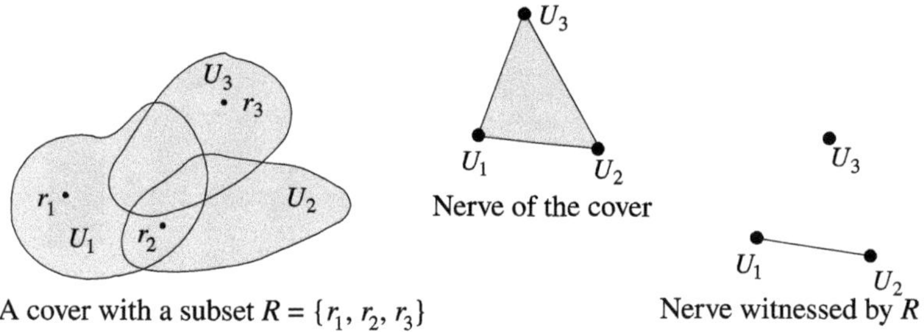

Fig. 4.24 An open cover $\mathscr{U}$ (*left*), its nerve $N(\mathscr{U})$ (*middle*), and its witnessed nerve $N(\mathscr{U}, R)$ (*right*)

Corollary 4.2 *If there is a receiver in each k-wise intersection of $\mathscr{U}$, then the cohomologies of the constant sheaves over $N(\mathscr{U})$ and $N(\mathscr{U}, R)$ are isomorphic. If there is not, then the cohomologies of $N(\mathscr{U})$ and $N(\mathscr{U}, R)$ may be different, as shown in Fig. 4.24.*

Algorithm 2 is fairly simple to implement, and can produce good results if the coverage regions aren't too complicated as the previous example shows. However, for poor signal coverage or inappropriate choice of thresholds, the nerve will not reflect the topology of the space, as shown in Fig. 4.23. Chapter 6 addresses the robustness of the nerve as thresholds are varied. In particular, Sect. 6.6 shows that this particular dataset is severely undersampled.

Corollary 4.2 provides some measure of control on the number of receivers or transmitters necessary to properly witness the nerve. The more substantial problem is that the coverage regions may intersect, but not in a contractible set. This problem is particularly acute in the case of undersampled data, because the threshold needed satisfy the Nerve Lemma will likely not be witnessed. (See Fig. 4.25 for an example.)

Increasing the detection threshold resolves this problem, but if there are enough measurements, randomly downsampling them can also be effective. For instance, using all measurements (on the rectangular 0.5 m spaced grid, a total of more than 2,500 receiver locations) at a detection threshold of -22 dB produced the nerve at right in Fig. 4.23, which fails to detect the courtyard. Downsampling to only 100

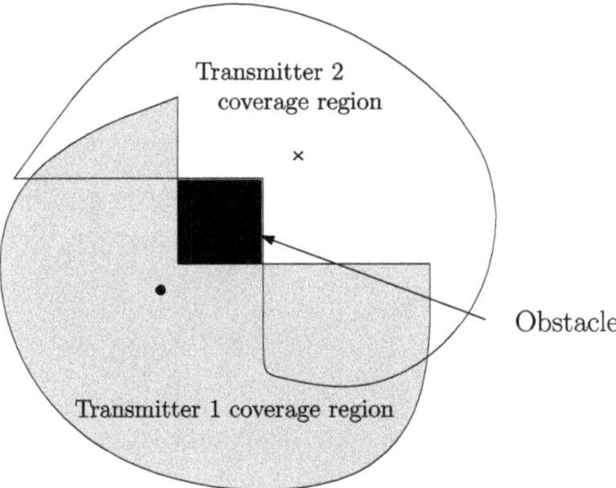

Fig. 4.25 A problem with the intersection of two coverage regions; it's not contractible

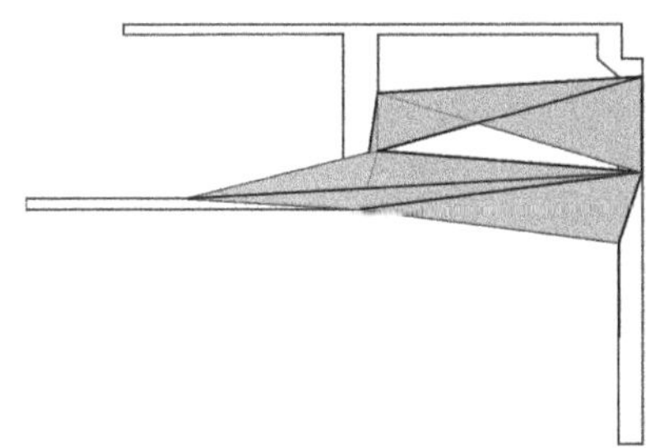

Fig. 4.26 Output of Algorithm 2 with witness threshold at 1 location and detection threshold at −22 dB, but using 100 randomly placed receivers

receiver locations recovers the missing topological feature representing the courtyard, as shown in Fig. 4.26.

Instead of assembling the nerve of the coverage regions for the transmitters, it is possible to build a nerve using the coverage regions for each receiver. Namely, the coverage regions $\mathcal{V} = \{V_j\}_{j=1}^{m}$ are assigned to each receiver r_j over which a transmitter will be detected by r_j. We could then construct $N(\mathcal{V}, T)$. If there are fewer receivers than transmitters, then $N(\mathcal{V}, T)$ will have fewer cells than $N(\mathcal{U}, R)$ which can ease the memory required for storing the nerve.

At first glance, it would seem that $N(\mathcal{U}, R)$ and $N(\mathcal{V}, T)$ are very different. However, both complexes come from a single relation $P \subset T \times R$ where $t_i P r_j$ if and only if r_j can detect a transmission from t_i. According to Dowker's theorem below, the geometric realizations $|N(\mathcal{U}, R)|$ and $|N(\mathcal{V}, T)|$ are homotopy equivalent.

Theorem 4.7 *(Dowker's theorem) Suppose that $P \subset T \times R$ is a relation between two finite sets. Suppose that K is an abstract simplicial complex in which each simplex is a collection of elements of R related to a single element $t \in T$. Dually, suppose that L is an abstract simplicial complex in which each simplex is a collection*

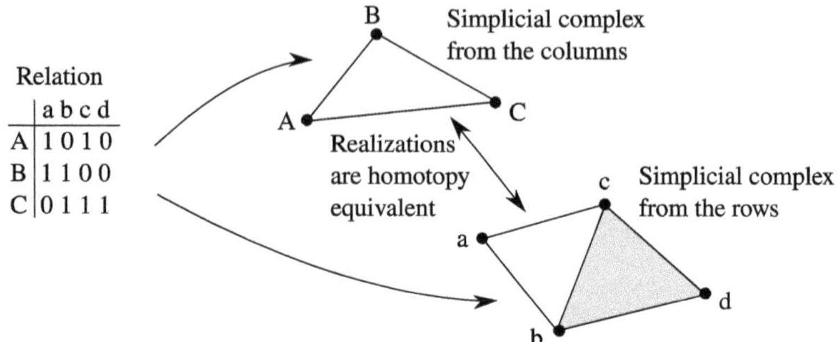

Fig. 4.27 An example of a relation and the two complexes constructed by Dowker's theorem

of elements of T related to a single element $r \in R$. Then the realizations of K and L are homotopy equivalent (See Fig. 4.27).

Proof (sketch; see Dowker (1952) for a complete proof) The idea of the proof is that cellular maps $f : K \to L$ and $g : L \to K$ can be constructed which take each simplex in K to a simplex in L (nonuniquely) and vice versa. (For technical reasons, it is sometimes necessary to subdivide a simplex first, thereby obtaining a cellular map from a refinement of the domain rather than K or L.) For instance, f is constructed so that every vertex of $f(\sigma)$ is R-related to every vertex of σ. This is possible by the constructions of K and L. Under the appropriate choice of maps, f and g form a homotopy equivalence. □

Example 4.22 Consider the relation shown in Fig. 4.27 and the associated complexes K (associated to the columns of R) and L (associated to the rows of R). From the figure, it is clear that the geometric realizations of the two complexes are homotopy equivalent, since it is possible to deform L into K after relabeling simplices.

Considering the maps f and g defined in the sketch of the proof, there is only one possibility for f. This is given by

$$f(\{A\}) = \{a\},\ f(\{B\}) = \{b\},\ f(\{C\}) = \{c\},$$
$$f(\{A, B\}) = \{a, b\},\ f(\{A, C\}) = \{a, c\},\ f(\{B, C\}) = \{b, c\}$$

Analogously, the only choice for g is

$$g(\{a\}) = \{A\},\ g(\{b\} = \{B\},\ g(\{c\}) = \{C\},\ g(\{d\}) = \{C\},$$
$$g(\{a, b\}) = \{A, B\},\ g(\{a, c\}) = \{A, C\},\ g(\{b, c\}) = \{B, C\},$$
$$g(\{b, d\}) = \{B, C\},\ g(\{c, d\}) = \{C\},\ g(\{b, c, d\}) = \{C\}.$$

4.8 Open Questions

1. The exact sequence of sheaf cohomology for a cellular pair (X, Y) appears to be a powerful inference tool. However, its use in the proof of the sampling theorems appears to require detailed knowledge about the sheaf in question. Is there a general sampling theorem that is less sensitive to the specific sheaf being sampled? For instance, can Propositions 4.2 and 4.3 be generalized?
2. Is there always a sharp bound on the sampling rate required to recover signals over networks (such as exhibited in Propositions 4.4 and 4.5), or are there classes of networks for which such bounds cannot be given?
3. The nerve computed using Algorithm 2 appears to be rather sensitive to noise. Can its robustness be improved, especially given that *reducing* the sampling rate can apparently improve the result?

References

Borsuk K (1948) On the imbedding of systems of compacta in simplicial complexes. Fund Math 35:217–234
Bott R, Tu L (1995) Differential forms in algebraic topology. Springer, New York
Bredon G (1997) Sheaf theory. Springer, Berlin
Dowker C (1952) Homology groups of relations. Ann Math 56(1952):84–95
Hatcher A (2002) Algebraic topology. Cambridge University Press, Cambridge
Nyquist H (1928) Certain topics in telegraph transmission theory. Trans AIEE 47:617–644
Shannon CE (1949) Communication in the presence of noise. Proc Inst Radio Eng 37(1):10–21
Shepard A (1980) A cellular description of the derived category of a stratified space. Ph.D. thesis, Brown University

Chapter 5
Transforms

This chapter will

1. Explain the Euler characteristic, a particular detector,
2. Connect the theory of detectors to the theory of integrals using the Euler characteristic,
3. Explain how the Euler integral counts targets whose shape is uncertain, and
4. Use the Euler characteristic to develop a class of hybrid geometric/topological transforms whose shape selectivity can be tailored.

Many filters arise from operating locally in a transformed space of functions. In contrast to filters, the transforms themselves are non-local and have global symmetries. For instance, the uncertainty principle in the Fourier transform ensures that the value of a function at a particular point can impact all frequencies. Most transforms in traditional signal processing are geometrically rigid, and are associated to a particular spatial geometry. For instance (Antimirov et al. 1993),

- Fourier series are associated to tori with definite radii,
- the Fourier transform is associated with Euclidean space,
- the Bessel transform is associated to disks,
- the spherical harmonics are associated to spheres, and
- the Mellin transform is associated with infinite wedges.

The usefulness of a transform relies on matching it to the domain; if the domain's geometry is unknown, then it is unclear which transform to use. Even if the domain is known, much effort has been expended by many researchers in developing approximate transforms, or modifying existing transforms for them. These are noble pursuits, resulting in useful matched filters that mitigate distortions and allow substantial improvements in signal-to-noise ratio. On the other hand, any deviation from the expected domain degrades the effectiveness of a transform-based method.

This chapter discusses transforms that use topological invariants to decrease their sensitivity to the domain. Although presently there is no purely topological transform theory, in which transforms are tailored to topological type rather than geometry, there

M. Robinson, *Topological Signal Processing*,
Mathematical Engineering, DOI: 10.1007/978-3-642-36104-3_5,

is a way to reduce the impact of certain geometric dependencies. By developing the theory of transforms based on the Euler integral, one can control which symmetries the resulting transform preserves. Unlike traditional integration theory, based on the Lebesgue measure, the Euler integral uses the Euler characteristic (a simple, but fundamental, topological invariant) in place of a measure. We discuss several important examples of Euler integral transforms and their potential uses in applications.

5.1 The Euler Characteristic

Suppose that

$$\cdots \xrightarrow{d_{k-1}} C_k \xrightarrow{d_k} C_{k+1} \xrightarrow{d_{k+1}} \cdots \tag{5.1}$$

is a chain complex (Definition 4.6).

Definition 5.1 The *Euler characteristic* of the chain complex $(C_\bullet, d_\bullet)$ is the alternating sum

$$\chi(C_\bullet) = \sum_k (-1)^k \dim C_k$$

of dimensions of the cochain groups.

If infinitely many terms of the chain complex are nontrivial, or any are infinite dimensional, then the Euler characteristic is not defined. It is clear that the Euler characteristics of two isomorphic chain complexes are the same. What is more surprising is that the Euler characteristic only depends on the homology of the chain complex, especially since the maps d_k do not appear in the definition of the Euler characteristic.

Example 5.1 Consider the chain complex (see Exercise 4.4)

$$0 \longrightarrow \mathbb{R}^4 \xrightarrow{d^0} \mathbb{R}^6 \xrightarrow{d^1} \mathbb{R}^4 \xrightarrow{d^2} \mathbb{R} \longrightarrow 0.$$

where

$$d^0 = \begin{pmatrix} 1 & -1 & 0 & 0 \\ 1 & 0 & -1 & 0 \\ 1 & 0 & 0 & -1 \\ 0 & 1 & -1 & 0 \\ 0 & 1 & 0 & -1 \\ 0 & 0 & 1 & -1 \end{pmatrix},$$

$$d^1 = \begin{pmatrix} 0 & 0 & 0 & 1 & -1 & 1 \\ 0 & -1 & 1 & 0 & 0 & -1 \\ 1 & 0 & -1 & 0 & 1 & 0 \\ -1 & 1 & 0 & -1 & 0 & 0 \end{pmatrix},$$

and

$$d^2 = \begin{pmatrix} 1 & 1 & 1 & 1 \end{pmatrix}.$$

The Euler characteristic of this chain complex is $4 - 6 + 4 - 1 = 1$.

Now, let us consider the homology of this chain complex. Observe that by row reducing d^0, we find that rank $d^0 = 3$, rank $d^1 = 3$, and rank $d^2 = 1$. Thus $\dim H^0 = 4 - 3 = 1$, $\dim H^1 = (6 - 3) - 3 = 0$, and $\dim H^2 = (4 - 1) - 3 = 0$. Hence, $\dim H^0 - \dim H^1 + \dim H^2 = 1$, which agrees with the Euler characteristic of the chain complex.

Lemma 5.1 *If the Euler characteristic of chain complex* $(C_\bullet, d_\bullet)$ *is defined, then the Euler characteristic of its homology, given by*

$$\chi(H_\bullet) = \sum_k (-1)^k \dim H_k$$

has the same value.

Proof (standard, see for instance Hatcher 2002; Curry et al. 2012) Since the homology $H_k = \ker d_k / \text{image}\, d_{k-1}$, then

$$\dim \ker d_k = \dim H_k + \dim \text{ image } d_{k-1}.$$

However, since $d_{k+1} \circ d_k = 0$,

$$\dim C_k = \dim \ker d_k + \dim \text{ image } d_{k-2}.$$

Therefore,

$$\dim C_k = \dim H_k + \dim \text{ image } d_{k-1} + \dim \text{ image } d_{k-2}.$$

Now observe that

$$\begin{aligned} \chi(C_\bullet) &= \sum_k (-1)^k \dim C_k \\ &= \sum_k (-1)^k (\dim H_k + \dim \text{ image } d_{k-1} + \dim \text{ image } d_{k-2}) \\ &= \sum_k (-1)^k \dim H_k + \sum (-1)^k (\dim \text{ image } d_{k-1} + \dim \text{ image } d_{k-2}) \\ &= \sum_k (-1)^k \dim H_k + (\dim \text{ image } d_0 + 0) - (\dim \text{ image } d_0 + \dim \text{ image } d_1) + \cdots \end{aligned}$$

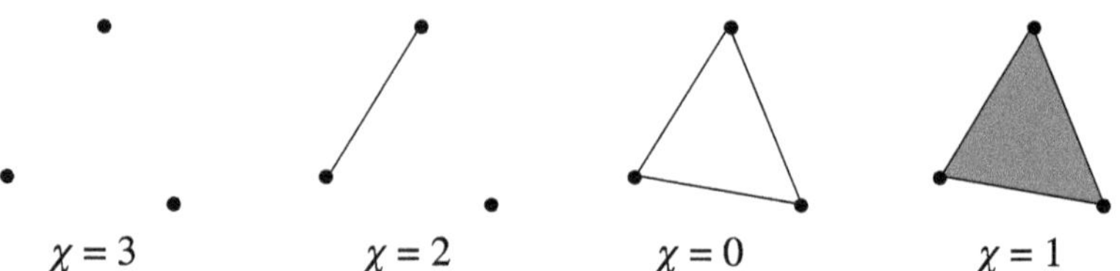

Fig. 5.1 The Euler characteristic generalizes the counting of connected components, though loops (*middle right*) destroy the analogy

$$= \sum_k (-1)^k \dim H_k.$$

□

Definition 5.2 (Dimca 2004) The *Euler characteristic* $\chi(\mathscr{S})$ of a sheaf $\mathscr{S}$ on X is the Euler characteristic of its cochain complex $(C^k(X;\mathscr{S}), d^k)$. The *compact Euler characteristic* $\chi_c(\mathscr{S})$ is the Euler characteristic of its compactly supported cochain complex $(C_c^k(X;\mathscr{S}), d^k)$. The *Euler characteristic of a cell complex X* is the compact Euler characteristic of the constant sheaf $\mathbb{R}$.

Proposition 5.1 *The Euler characteristic of a cell complex X is given by*

$$\chi_c(X) = \sum_{c\in X} (-1)^{\dim c} = \sum_{k=0}^{\dim X} (-1)^k \#\{\textit{cells of dimension } k\}.$$

Proof Simply construct the cochain complex for the constant sheaf $\mathbb{R}$. □

Example 5.2 If a CW complex is homotopy equivalent to a set of points, then its Euler characteristic is the number of its connected components (see the far left, middle left, and far right frames of Fig. 5.1). However, if the CW complex contains loops (middle right frame of Fig. 5.1), then the Euler characteristic will differ from the number of connected components. Specifically, if there are only vertices, the Euler characteristic is the number of vertices. Each edge has two endpoints, and so adding an edge usually removes a connected component. A loop has Euler characteristic zero, so each 2-dimensional cell always adds back an additional connected component by filling in a hole. As the definition of the Euler characteristic makes clear, this pattern continues for higher-dimensional cells.

Exercise 5.1 Table 5.1 exhibits the Euler characteristic of some familiar cell complexes, computed using Proposition 5.1.

1. Verify the entries in the table by constructing the cell complex and counting cells.
2. For each cell complex listed in the table, construct a different cell complex that is homotopy equivalent to it. What are the Euler characteristics of the complexes that you constructed?

Table 5.1 Euler characteristic of some familiar cell complexes, each with an assumed cell complex structure

Cell complex	dim C^0	dim C^1	dim C^2	χ
Interval (0, 1)	0	1	0	−1
Interval [0, 1)	1	1	0	0
Interval [0, 1]	2	1	0	1
Disk	1	1	1	1
Annulus	2	3	1	0
Cube	8	12	6	2
Tetrahedron	4	6	4	2
Torus	1	2	1	0

Proposition 5.2 *The cohomology of the constant sheaf is a homotopy invariant, and therefore does not depend on the particular choice of cell decomposition.*

For a proof of this proposition, see Hatcher (2002, Thm 2.27).

5.1.1 Valuations

The Euler characteristic generalizes counting for finite sets: if A is a finite set, then the Euler characteristic A is equal to the cardinality of A. Valuations also generalize counting, and the Euler characteristic is one of the two best-known valuations. Valuations are important because the concept of integration of functions extends to valuations; this leads to topologically-motivated integral transforms.

Definition 5.3 Suppose that $\mathscr{B}$ is a collection of sets that is closed under union and intersection. A *valuation* on $\mathscr{B}$ is a function $v : \mathscr{B} \to \mathbb{R}$ that satisfies

$$v(A \cup B) = v(A) + v(B) - v(A \cap B) \tag{5.2}$$

for all $A, B \in \mathscr{B}$.

Example 5.3 The simplest example of a valuation is the *counting measure* $|\cdot|$, which returns the cardinality of a set. The counting measure is defined on the collection of all sets, in which case (5.2) is merely the inclusion-exclusion principle.

Valuations therefore generalize the concept of counting. They can also describe counting only the sets that "matter" as the next exercise shows.

Exercise 5.2 Suppose that $\mathscr{B}$ is closed under union and intersection and that $x \in \bigcup \mathscr{B}$. Show that the *Dirac measure δ_x concentrated at x* defined by the formula

$$\delta_x(A) = \begin{cases} 1 \text{ if } x \in A \\ 0 \text{ otherwise.} \end{cases}$$

is a valuation.

Valuations also can be used for measuring area. Suppose that $\mathscr{B}$ is a finite collection of open subsets of $\mathbb{R}^2$ that is closed under unions and intersections. Then for a set $A \in \mathscr{B}$, let

$$a(A) = \iint_A dx\, dy.$$

Exercise 5.3 Using the properties of integrals, show that a is a valuation.

Remark 5.1 Readers familiar with the definition of a measure will recognize that all measures are valuations. However, there are two ways that valuations may fail to be measures, namely

1. measures of a set must be nonnegative, but valuations can take negative values, and
2. measures are σ-additive, while valuations usually are not.

The lack of σ-additivity typically means that all calculations involving valuations must be finite; this is why cell complex decompositions and constructible functions play an important role in the following sections. Since measures are somewhat better-behaved, one can make practical use of them on substantially more pathological collections of sets than is feasible with valuations.

Proposition 5.3 *Suppose that X is a cell complex and that $A \cup B = X$ are two subcomplexes. Then $\chi(X) = \chi(A) + \chi(B) - \chi(A \cap B)$, so the Euler characteristic is a valuation.*

Proof Observe that for the constant sheaf $\mathscr{L}$, the following sequence of cochains

$$0 \longrightarrow C_c^k(X; \mathscr{L}) \longrightarrow C_c^k(A; \mathscr{L}) \oplus C_c^k(B; \mathscr{L}) \longrightarrow C_c^k(A \cap B; \mathscr{L}) \longrightarrow 0$$

is exact. (This is the Mayer-Vietoris short exact sequence for constant sheaves.) This means that

$$\dim C_c^k(X; \mathscr{L}) = \dim C_c^k(A; \mathscr{L}) + \dim C_c^k(B; \mathscr{L}) - \dim C_c^k(A \cap B; \mathscr{L})$$

whence the result follows immediately. □

5.1.2 The Euler Integral

In order to build a suitable integration theory based on valuations, the functions we integrate should be suitably controlled. Following the common theme of this book, the appropriate functions are those that are compatible with an underlying cellular structure.

Definition 5.4 A function $f : X \to Y$ between topological spaces is *constructible* if there exists a (nonunique) cell complex X_f for which (a) there exists a homeomorphism

$h : X_f \to X$ and (b) $f \circ h$ is constant on each cell of X_f (Van Den Dries 1998). (Constructible functions are usually *not* continuous.) The space of constructible functions from X to Y is denoted $CF(X, Y)$. We will usually abuse notation and write f instead of $f \circ h$.

Constructible functions are a particularly nice generalization of piecewise constant functions, in which the discontinuities are confined to occur at cell boundaries. Without too much loss of generality, we therefore usually constrain our attention to $CF(X, \mathbb{Z})$, *integer-valued* constructible functions.

Remark 5.2 In this book, we have been inspired by Shepard (1980) to use cellular sheaves, which respect a fixed cell complex structure. These are very effective tools for signal processing, but their use is at variance with the traditional literature on sheaves. Under the more traditional definition, $CF(X, Y)$ is a sheaf (Schürmann 2003). Since each function in $CF(X, Y)$ is constructible with respect to a different cell complex structure, $CF(X, Y)$ is not a cellular sheaf. This is a weakness of cellular sheaves, though it does not impact our exposition here.

Valuations are important because they permit the concept of integration to be extended to functions of cell complexes in a rather general way.

Definition 5.5 Suppose $f \in CF(X, \mathbb{Z})$, and that v is a valuation on the cells of X_f. Then the v-integral of f is

$$\int f \, dv = \sum_{k=-\infty}^{\infty} kv\left(f^{-1}(k)\right).$$

If $v = \chi$, the Euler characteristic, then the resulting integral is called the *Euler integral*.

Example 5.4 Computation of the Euler integral of a constructible function can be subtle. Figure 5.2 shows a constructible function and the breakdown of its Euler integral computation. It is essential to account for the boundaries of each level set correctly in order to arrive at the correct answer. In this example, we assume that the function is upper semicontinuous, which means that the value of the function at each jump discontinuity is equal to the largest of its possible limits. Therefore, the level set where the function takes the value 4 includes its boundary. The other level sets do not include the inner boundaries, but do include the outer boundaries.

Exercise 5.4 1. Show that if $f \in CF(X, \mathbb{Z})$ only takes on finitely many values, then

$$\int f \, dv = \sum_{k=0}^{\infty} v\left(f^{-1}((k, \infty))\right) - v\left(f^{-1}((-\infty, -k))\right).$$

2. This form of the Euler integral is more tolerant to discretization error than computing the Euler characteristic of level sets (Baryshnikov and Ghrist 2009). Explain

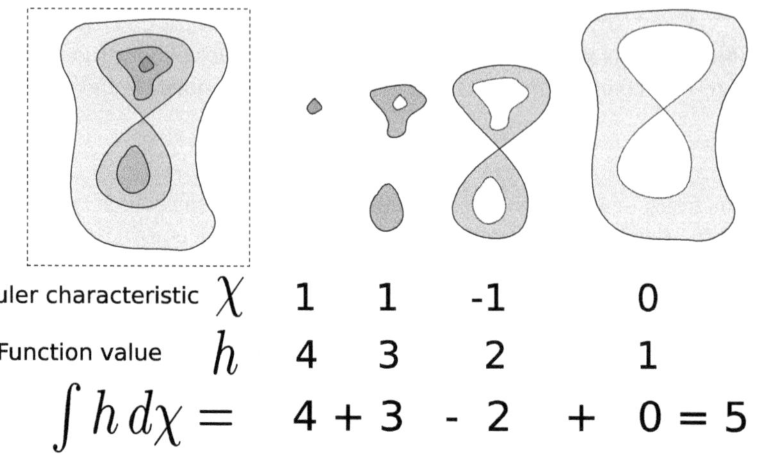

Fig. 5.2 An example Euler integral computation

the difference between the two formulas for computing $\int f\ dv$ when f is evaluated at a discrete grid of points.

Example 5.5 The v-integrals of constructible indicator functions are particularly simple. Suppose that $A \subset X$ is a subcomplex of a cell complex X and that 1_A is the indicator function on A, namely that

$$1_A(x) = \begin{cases} 1 \text{ if } x \in A \\ 0 \text{ otherwise} \end{cases}$$

Using the definition of the v-integral,

$$\int 1_A\ dv = 1v\left(1_A^{-1}(1)\right) = v(A).$$

So the v-integral is a tool for extending valuations from sets to functions on sets.

Proposition 5.4 *The v-integral, as a function $CF(X;\mathbb{Z}) \to \mathbb{Z}$ is linear.*

Proof Suppose that $h, g \in CF(X;\mathbb{Z})$. By definition, this means that they can be written in the form of sums of indicator functions

$$h = \sum_j a_j 1_{A_j},\ g = \sum_k b_k 1_{B_k}.$$

Thus

$$\begin{aligned} h+g &= \sum_j a_j 1_{A_j} + \sum_k b_k 1_{B_k} \\ &= \sum_j \sum_k (a_j + b_k) 1_{A_j \cap B_k}. \end{aligned}$$

Considering the integral, we obtain

$$\begin{aligned} \int h+g \, d\nu &= \int \sum_j \sum_k (a_j + b_k) 1_{A_j \cap B_k} d\nu \\ &= \sum_j \sum_k (a_j + b_k) \int 1_{A_j \cap B_k} d\nu \\ &= \sum_j \sum_k (a_j + b_k) \nu(A_j \cap B_k) \\ &= \sum_j a_j \nu(A_j) + \sum_k b_k \nu(B_k) \\ &= \int h d\nu + \int g d\nu. \end{aligned}$$

□

Example 5.6 Like integration against a measure, ν-integration generalizes the idea of a weighted sum. Suppose that A is a finite set. Then if h is a constructible $\mathbb{Z}$-valued function on A, then

$$\begin{aligned} \int h \, d\nu &= \sum_k k\nu(h^{-1}(k)) \\ &= \sum_k k \sum_{\{a \in A : h(a) = k\}} \nu(\{a\}) \\ &= \sum_{a \in A} h(a) \nu(\{a\}) \end{aligned}$$

which is a weighted sum over A.

Remark 5.3 The definition of a ν-integral seems to give substantial flexibility in the kinds of integrals that can be defined. However, we usually want an integral to have certain invariance properties, which restricts the possiblities for the valuation. If we require a valuation to be dilation-, translation-, and rotation-invariant, then a statement called Hadwiger's theorem (see for instance Chen (2004)) implies there is exactly one choice: a linear scaling of the Euler characteristic. This indicates that any transform of functions defined by integrating against a valuation that is "topological" must use the Euler integral.

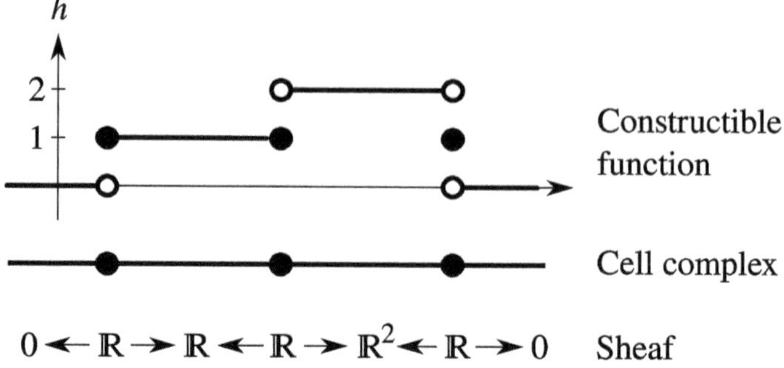

Fig. 5.3 Encoding a constructible function $h \in CF(X, \mathbb{Z})$ as a sheaf $\mathscr{H}$

The Euler integral has an interpretation in terms of compact Euler characteristics of sheaves (Kashiwara and Schapira 1990, [Sect. 9.1]; Dimca 2004). Suppose that $h \in CF(X; \mathbb{Z})$ is a nonnegative function. It is easy to construct a sheaf $\mathscr{H}$ whose stalk at x has dimension $h(x)$, as shown in Fig. 5.3. While not unique, $\mathscr{H}$ can be constructed as the sum (Definition 3.8) of constant sheaves over subcomplexes of X. Observe that

$$\begin{aligned}\chi_c(\mathscr{H}) &= \sum_k (-1)^k \dim H_c^k(X; \mathscr{H}) \\ &= \sum_k (-1)^k \dim C_c^k(X; \mathscr{H}) \\ &= \sum_k (-1)^k \sum_{x \in X^k} h(x) \\ &= \sum_k \sum_{x \in X^k} (-1)^k h(x).\end{aligned}$$

Since the double sum ranges over all cells in X, we can compute this sum by ranging over level sets of h instead, to obtain

$$\begin{aligned}\chi_c(\mathscr{H}) &= \sum_{s=0}^{\infty} s \sum_{x \in X : h(x) = s} (-1)^{\dim x} \\ &= \sum_{s=0}^{\infty} s \chi(h^{-1}(s)) \\ &= \int h \, d\chi.\end{aligned}$$

This derivation relied the assumption that h was nonnegative since it was used as a dimension of the stalks. This restriction can be lifted by instead considering sequences of sheaves.

Definition 5.6 Suppose that

$$\cdots \longrightarrow \mathscr{S}_k \longrightarrow \mathscr{S}_{k+1} \longrightarrow \cdots$$

is a sequence of sheaves. The *compact Euler characteristic* $\chi_c(\mathscr{S}_\bullet)$ of this sequence is given by

$$\chi_c(\mathscr{S}_\bullet) = \sum_{k=-\infty}^{\infty} (-1)^k \chi_c(\mathscr{S}_k).$$

Example 5.7 Given that we represent a constructible function h by the two-term sequence

$$0 \longrightarrow \mathscr{H}_0 \longrightarrow \mathscr{H}_1 \longrightarrow 0,$$

where

$$\dim \mathscr{H}_0(x) = \begin{cases} h(x) \text{ if } h(x) \geq 0 \\ 0 \quad\quad \text{otherwise} \end{cases}$$

and

$$\dim \mathscr{H}_1(x) = \begin{cases} -h(x) \text{ if } h(x) \leq 0 \\ 0 \quad\quad\quad \text{otherwise} \end{cases}$$

the previous calculation immediately shows that $\chi_c(\mathscr{H}_\bullet) = \int h \, d\chi$.

5.2 Case Study: Target Enumeration

Because the Euler characteristic is an extension of counting (Example 5.2), the Euler integral can be used to enumerate certain features of a signal. Consider the task of counting a number of target signals using a dense field of sensors, each of which returns a count of these target signals in its vicinity. Suppose that each sensor does not assign a unique ID to each target signal it detects. Surprisingly, under mild topological assumptions about the signals, the number of targets can be recovered.

Although dense fields of sensors are not particularly common in practice, they do arise in a number of important settings, such as

1. The array of sensors of a digital camera (in which they are arrayed in a 2-dimensional rectangular grid),
2. An antenna array (usually 1- or 2-dimensional), in which each antenna element is addressed separately,
3. In a 3-dimensional collection of underwater microphones (such as those used for tracking the migration patterns of whales),
4. In configurations that blend spatial sensor diversity with time diversity, such as

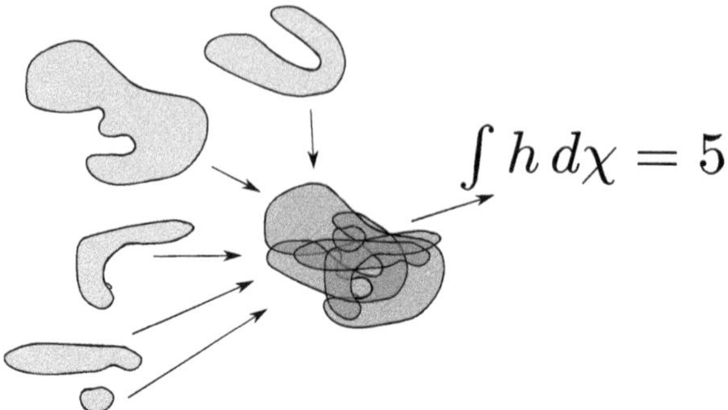

Fig. 5.4 Counting the number of overlapping sets via the Euler integral

(a) A single RFID interrogating sensor, which operates as a 1-dimensional (in time) network of sensors
(b) A traffic counter, counting the number of vehicles passing over its hose
(c) A synthetic aperture imaging sensor, again as a 1-dimensional sensor network
(d) A transversely-scanning sonar sensor mounted on a moving vessel, which operates as a 1-dimensional spatial network (in angle) plus a 1-dimensional network in time.

We axiomize the target signals in a topological space X in which the sensors and targets are located. To target k, assign an open set $U_k \subseteq X$ called the *target support* that consists of all locations at which a sensor detects the target. However, because we assumed that target identities cannot be reported by the sensors, the most information that could ever be available to a sensor network is the *height function* $h : X \to \mathbb{Z}$ which is defined by letting $h(x)$ be the number of elements of $U_k \in \mathscr{U}$ that contain $x \in X$. Of course, the sensor network will return a discrete sampling of this function at a finite number of points $\{x_1, \dots, x_n\} \subset X$.

The following calculation due to Baryshnikov and Ghrist (2009) shows how to recover the number of elements of $\mathscr{U} = \{U_1, \dots\}$ from h (see Fig. 5.4). It is useful to think of h as the sum of indicator functions

$$h(x) = \sum_k 1_{U_k}.$$

Suppose that $U_k \in \mathscr{U}$, then

$$\int h\, d\chi = \int \sum_k 1_{U_k} d\chi = \sum_k \int 1_{U_k} d\chi = \sum_k \chi(U_k) = |\mathscr{U}| \chi(U_k).$$

Provided $\chi(U_k) \neq 0$, then

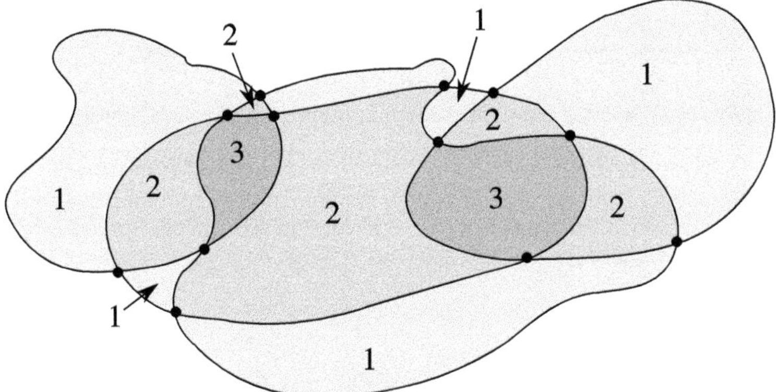

Fig. 5.5 Values of the function h in which four target support sets overlap

$$|\mathscr{U}| = \frac{1}{\chi(U_k)} \int h \, d\chi. \tag{5.3}$$

The interpretation of this calculation is somewhat striking. Suppose that there is a finite collection of large targets that are visible to a dense field of sensors spread over a region X. If each sensor returns an anonymous count of the targets visible to it, this calculation shows how to recover the total number of targets from these counts.

Example 5.8 Consider the function h shown in Fig. 5.5, which arises from the collection of four contractible target support sets. (Since each target support is an open set, this function is lower semicontinuous; the value along each boundary is the lower of the two neighboring values.) Tabulating the values of the function according to the dimension of the cells, we have

Value	0-cells	1-cells	2-cells	$\chi(h^{-1}(\{\text{value}\}))$
1	6	12	6	0
2	0	6	5	−1
3	0	0	2	2

Accumulating these values, the Euler integral is $3 \times 2 + 2 \times (-1) = 4$, which is the number of target support sets.

The example above embodies the idealized setting in which the field of sensors always returns correct counts and is continuous. However, both of these sources of error are far from benign. Since each vertex has Euler characteristic 1, the Euler integral is quite sensitive to errors. The treatment of errors due to random perturbations of the sensor output is quite technical and is largely unexplored; the interested reader should see Bobrowski and Borman (2012).

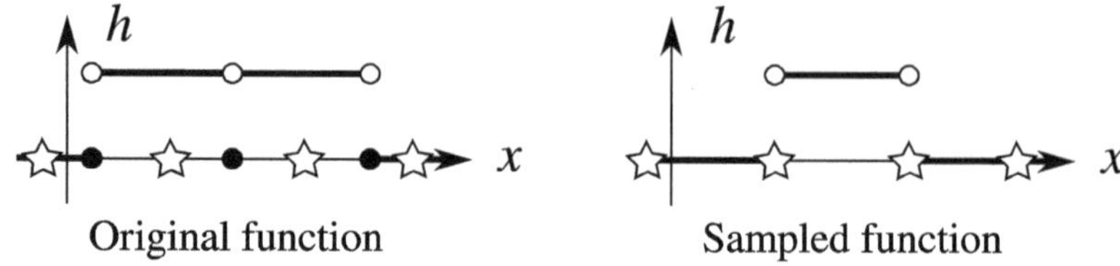

Fig. 5.6 The function at left consists of two target support sets and has Euler integral −2. The function at right is sampled from the one at left using the points marked with *stars*, and has Euler integral −1

Discretization of a constructible function is a delicate operation, since constructible functions imply a particular cell complex decomposition of the underlying space. If there are only finitely many sensors, though, it is necessary to discretize the Euler integral. If the geometry of the topological space X is known, it is usually most effective to let the sensors be the vertices in a Delaunay tesselation of X. If the geometry is not known, then one way to proceed is to construct the Vietoris-Rips complex, described in Sect. 6.2. Once the discretizaton of X has been specified, the sensor values can be extended to a height function over all cells of X by way of lower-semicontinuity. Specifically, the value at a given cell is the minimum value over all (higher dimensional) cells that it attaches to.

Any discretization procedure is subject to errors when target support sets (or the level sets of their height function) are tangent or close to being tangent. In this case, it is difficult to determine if two sets are connected or not. This topological uncertainty results in an inaccurate estimate of the Euler integral, which degrades its value in target enumeration. Although in some situations (see Krupa (2012)) it is possible to compensate for the bias that this introduces using statistical estimates, the only true remedy appears to be a high sensor density.

Example 5.9 Consider the function shown at left in Fig. 5.6. Recall that the Euler characteristic of an open interval is -1, so the Euler integral is negative the number of targets. It is immediate that the Euler integral of this function is -2, which correctly recovers the number of targets. However, if only the four points marked with stars can be observed, then the estimate for the function is shown at right. This function has Euler integral -1, which is incorrect.

5.3 Euler Integral Transforms

The study of Euler integral transforms began with a paper by Schapira (1995). This chapter was ahead of its time, and the ideas lay dormant until a flurry of recent activity (Curry et al. 2012; Ghrist and Robinson 2011; Baryshnikov et al. 2011). All of the Euler integral transforms that have been developed have certain invariances that allow a trade off between topological flexibility and geometric rigidity.

Definition 5.7 Suppose that $L : C(M) \to C(N)$ is a linear transformation of continuous functions on a manifold M to continuous functions on a manifold N. Observe that any diffeomorphism $d : M \to M$ induces a linear map $\tilde{d} : C(M) \to C(M)$ on the space of continuous functions. A pair of diffeomorphisms $d : M \to M$ and $e : N \to N$ is called an *invariance* for L if the following diagram commutes

$$\begin{array}{ccc} C(M) & \xrightarrow{L} & C(N) \\ \downarrow{\tilde{d}} & & \downarrow{\tilde{e}} \\ C(M) & \xrightarrow{L} & C(N) \end{array}$$

Example 5.10 Observe that the Euler integral is a linear operation because of Proposition 5.4. If M is a manifold, then Proposition 5.2 implies that any diffeomorphism $M \to M$ will not change the Euler characteristic. Hence the Euler integral is invariant under any diffeomorphism paired with the identity.

Exercise 5.5 Show that the transformation $(x\ \ y)^T \mapsto (x'\ \ y')^T$ given by

$$\begin{pmatrix} x' \\ y' \end{pmatrix} = \begin{pmatrix} x\cos\theta - y\sin\theta + p \\ x\sin\theta + y\cos\theta + q \end{pmatrix}.$$

is an invariance when paired with the identity for the Laplacian operator

$$\frac{\partial^2}{\partial x^2} + \frac{\partial^2}{\partial y^2}.$$

Definition 5.8 (Compare to Schapira 1995; Baryshnikov et al. 2011, which treats the Radon transform in a similar way.) Suppose that M and N are manifolds and that $P : M \times N \to \mathbb{R}$ is a smooth function. An *Euler integral transform* $\mathbf{T}$ *subordinate to* P takes each constructible $f \in CF(M, \mathbb{Z})$, to a function $\mathbf{T}f : N \to \mathbb{R}$ given by the formula

$$\mathbf{T}f(x) = \int_{-\infty}^{\infty} k_1(x, r) \int_{\{y \in M : P(y,x) = r\}} k_2(x, y) f(y) d\chi(y)\, dr$$

where $k_1 : N \times \mathbb{R} \to \mathbb{R}$ is smooth and $k_2(x, \cdot) \in CF(M, \mathbb{Z})$. Each surface $\{y \in M : P(y, x) = r\}$ is called an *isocline* for the transform (see Fig. 5.7).

Remark 5.4 Observe that if we instead consider a Lebesgue integral as the inner integral, the Fubini theorem applies so that

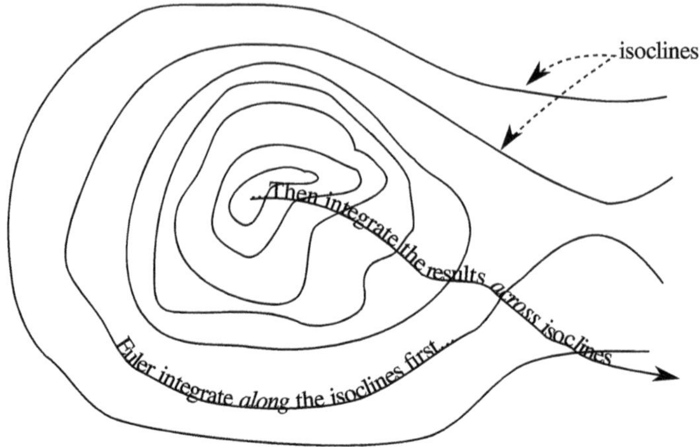

Fig. 5.7 An Euler integral transform operates by integrating along a collection of isoclines using the Euler characteristic, and then integrating these results using the Riemann or Lebesgue integral

$$\begin{aligned}
\mathbf{T}f(x) &= \int_{-\infty}^{\infty} k_1(x,r) \int_{\{y\in M:P(y,x)=r\}} k_2(x,y)f(y)dy\,dr \\
&= \int_{-\infty}^{\infty} \int_{\{y\in M:P(y,x)=r\}} k_1(x,r)k_2(x,y)f(y)dy\,dr \\
&= \int_M k_1(x,P(y,x))k_2(x,y)f(y)dy \\
&= \int_M k(x,y)f(y)dy,
\end{aligned}$$

which is the typical formula for an integral transform from functions on M to functions on N. Note that this derivation does not work for the hybrid combination of Lebesgue and Euler integration, so we must treat the Euler and Lebesgue integrals separately.

Remark 5.5 Euler integral transforms cannot be morphisms of sheaves, because they are manifestly not local from the definition. The value of a transformed function $\mathbf{T}f$ at $x \in N$ depends on values of the function f at *all* $y \in M$.

The control of the invariances of an Euler integral is an important motivation for their use; the following proposition summarizes how invariances may be identified.

Proposition 5.5 *A pair of diffeomorphisms $D_M : M \to M$ and $D_N : N \to N$ are an invariance for an Euler integral transform if*

1. $P(y, D_N x) = P(D_M y, x)$ *for all* $x \in N$ *and* $y \in M$,
2. $k_1(D_N x, r) = k_1(x, r)$ *for all* $x \in N$ *and* $r \in \mathbb{R}$, *and*
3. $k_2(D_N x, y) = k_2(x, D_M y)$ *for all* $x \in N$ *and* $y \in M$.

Proof We merely proceed by direct computation:

$$\begin{aligned}
(D_N \circ \mathbf{T})f(x) &= \mathbf{T}f(D_N x) \\
&= \int_{-\infty}^{\infty} k_1(D_N x, r) \int_{\{y \in M : P(y, D_N x) = r\}} k_2(D_N x, y) f(y) d\chi(y)\, dr \\
&= \int_{-\infty}^{\infty} k_1(x, r) \int_{\{y \in M : P(D_M y, x) = r\}} k_2(x, D_M y) f(y) d\chi(y)\, dr \\
&= \int_{-\infty}^{\infty} k_1(x, r) \int_{\{y \in M : P(y, x) = r\}} k_2(x, y) f(D_M^{-1} y) d\chi(y)\, dr \\
&= \mathbf{T}(f \circ D_M^{-1})(x).
\end{aligned}$$

□

The remaining portion of this section examines two Euler integral transforms, in which the level sets are affine subspaces (the Euler-Fourier transform) or topological spheres (the Euler-Bessel transform) in $\mathbb{R}^n$.

5.3.1 The Euler–Fourier Transform

By analogy with the Fourier transform, which has constant phase along parallel planes, the Euler integral transform that uses parallel planes as isoclines is called the *Euler-Fourier transform.*

Definition 5.9 (See Ghrist and Robinson 2011; Curry et al. 2012) Suppose that $h \in CF(\mathbb{R}^n; \mathbb{Z})$. The *Euler-Fourier transform* of h is given by

$$\mathbf{EF}h(\xi) = \int_{-\infty}^{\infty} \int_{<\xi, x>=s} h(x) d\chi(x)\, ds,$$

where ξ is a unit vector $S^{n-1} \subset \mathbb{R}^n$, and $< \cdot, \cdot >$ is the usual dot product. The transform is shown schematically in Fig. 5.8.

Proposition 5.6 *The Euler-Fourier transform is rotation invariant. The Euler-Fourier transform is insensitive to translations.*

Proof Observe that if R is a rotation of $\mathbb{R}^n$, then $< R\xi, x >=< \xi, R^T x >$. So the Euler-Fourier transform is rotation invariant by Proposition 5.5. Observe that translation in x parallel to ξ does not change the value of $\mathbf{EF}h(\xi)$ because of the translation invariance of the Lebesgue integral. Similarly, translation perpendicular to ξ does not change the value of the transform because of the homeomorphism invariance of the Euler integral. □

The Euler-Fourier transform is sensitive to dilation, however. If h is the indicator function on a disk D, the Euler-Fourier transform is a constant function, equal to the diameter of D.

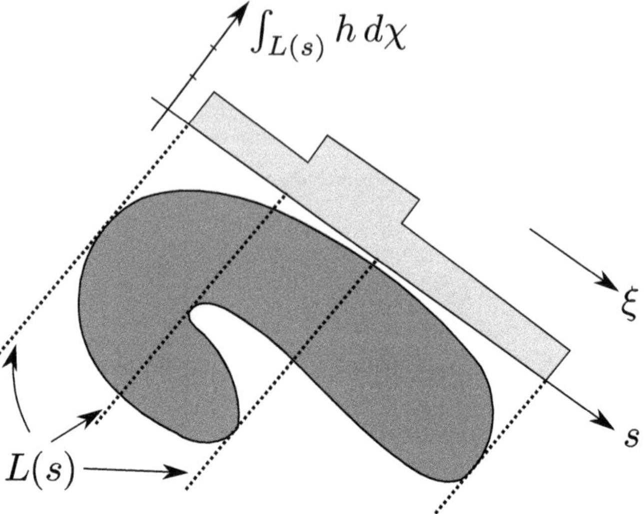

Fig. 5.8 Euler-Fourier transform of an indicator function on a set; the family of lines $L(s)$ are perpendicular to ξ

Consider the Euler-Fourier transform of the indicator function 1_L on a line segment L in the plane. The transform can be written explicitly in this case. Since the Euler-Fourier transform is rotation invariant, without loss of generality, we can assume that its endpoints are $(0, 0)$ and $(0, L)$. Observe that if $\xi = (\cos\theta, \sin\theta)$, then

$$\begin{aligned}\mathbf{EF}1_L(\xi) &= \int_{-\infty}^{\infty}\int_{\{<\xi,x>=s\}} 1_L(x)d\chi\ ds \\ &= \int_{-\infty}^{\infty} 1_{\{0\leq s\leq L\sin|\theta|\}}(s)ds \\ &= L\sin|\theta|.\end{aligned}$$

Since the Euler-Fourier transform is linear, we can use the transform of a line segment to compute the transforms of other polygons in the plane, merely by adding indicator functions on each side. Care is *not* needed at the vertices, since the Euler-Fourier transform of a half-open line segment is the same as the Euler-Fourier transform of a closed line segment. Figure 5.9 shows the graphs of the Euler-Fourier transforms of several regular polygons.

Corollary 5.1 *The Euler-Fourier transform is* not *a homeomorphism invariant, since the transforms of non-congruent polygons differ.*

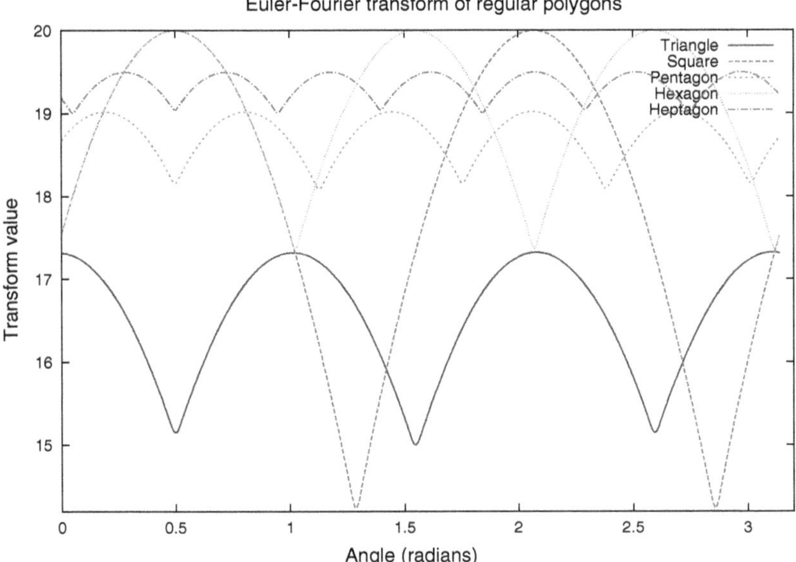

Fig. 5.9 Euler-Fourier transform of indicator functions on several *regular polygons*

5.3.2 Euler–Bessel Transform

We can step back from the translation insensitivity of the Euler-Fourier transform by making a small modification. It is reasonable to think of a set of parallel planes as being a set of concentric spheres whose centers are located at infinity (the "far field"). This idea connects the usual Fourier transform to the Bessel transform. Therefore, by analogy, we call the Euler integral transform which has spherical isoclines the *Euler-Bessel* transform.

Definition 5.10 (See Ghrist and Robinson 2011; Curry et al. 2012) Suppose that $\|\cdot\|$ is a norm on $\mathbb{R}^n$ and $h \in CF(X;\mathbb{Z})$. The *Euler-Bessel transform associated to* $\|\cdot\|$ of h is given by

$$\mathbf{EB}h(x) = \int_0^\infty \int_{\|x-y\|=s} h(y) d\chi(y)\, ds.$$

We usually call the Euler-Bessel transform associated to the usual 2-norm the *circular Euler-Bessel transform* **CEB** (see Fig. 5.10) and the Euler-Bessel transform associated to the ∞-norm the *square Euler-Bessel transform* **SEB**. These play an important role in our exposition, largely because they are easiest to compute. The many different kinds of Euler-Bessel transforms have different geometric properties, though they share the following invariances.

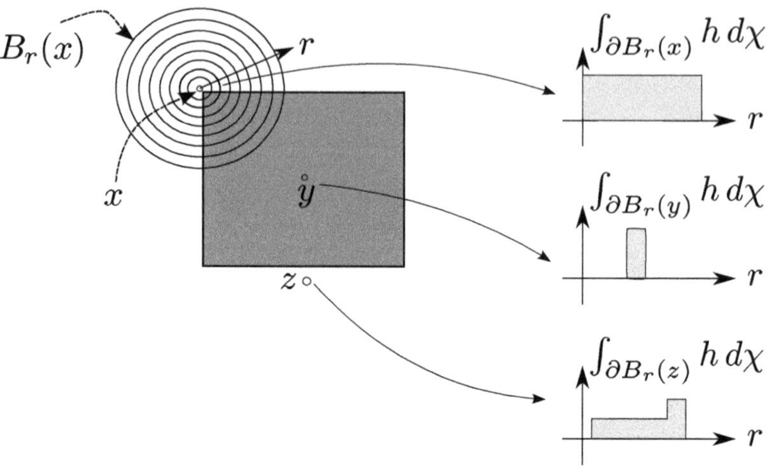

Fig. 5.10 Circular Euler-Bessel transform of an indicator function on a *square*; several representative values are shown

Theorem 5.1 *The Euler-Bessel transform is translation invariant, but unlike the Euler-Fourier transform only the circular Euler-Bessel transform is rotation -invariant.*

Proof The translation invariance is easy to see from Proposition 5.5 because $\|(x+h)-y\| = \|x-(y-h)\|$. Rotation invariance for the 2-norm also follows from Proposition 5.5. That calculation relies on the fact that

$$\begin{aligned}\|Rx-y\| &= \sqrt{<Rx-y, Rx-y>}\\ &= \sqrt{<Rx, Rx> -2 <Rx, y> + <y, y>}\\ &= \sqrt{<x, x> -2 <x, R^T y> + <R^T y, R^T y>}\\ &= \|x-R^T y\|.\end{aligned}$$

This is simply untrue for other norms. □

The circular Euler-Bessel transform of the indicator function on a disk has a striking form, as shown in Fig. 5.11. There is a local minimum, at the center of the disk. Outside the disk, the transform's value is equal to the diameter of the disk.

Using the same idea as for the Euler-Fourier transform, we can compute the transform of indicator functions on polygons by first treating the case of a line segment. The choice of norm matters in the final answer, unlike the Euler-Fourier transform.

Since the circular Euler-Bessel transform is rotation-invariant, we treat that first, by considering the line segment L from $A=(0,0)$ to $B=(0,L)$ as before. The key computation is then

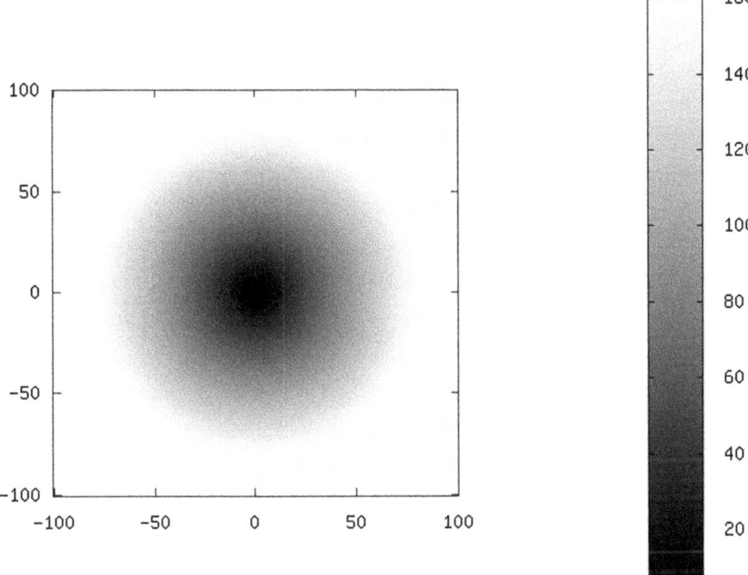

Fig. 5.11 Circular Euler-Bessel transform of an indicator function on a disk with radius 80 units

$$\mathbf{CEB}1_L(x) = \int_0^\infty \int_{\|x-y\|=s} 1_L(y) d\chi(y) \, ds,$$

where we observe that $I(y) = \int_{\|x-y\|=s} 1_L(y) d\chi(y)$ is the number of intersections between the circle of radius s centered at x and L. Figure 5.12 shows $I(y)$ for a particular fixed s. Integrating this function with respect to s yields the following explicit formula

$$\mathbf{CEB}1_L(x) = |d(A,x) - d(B,x)| + 2\left(\min\{d(A,x), d(B,x)\} - d(L,x)\right). \quad (5.4)$$

Exercise 5.6 The circular Euler-Bessel transform of a line segment is rather different from its square Euler-Bessel transform. The resulting transform of a vertical line segment is shown schematically in Fig. 5.13. Compute the square Euler-Bessel transform of a line segment that is not aligned with the cardinal axes.

Using the same reasoning as in the Euler-Fourier transform case, one can easily add the explicit formulas for the Euler-Bessel transforms of line segments to obtain expressions for polygons. For instance, Fig. 5.14 shows the transform of the indicator functions on several polygons.

From Fig. 5.14, it is visually apparent that the Euler-Bessel transform has local minima near the centers of each polygon. In the case of a circular Euler-Bessel transform, this is easily shown.

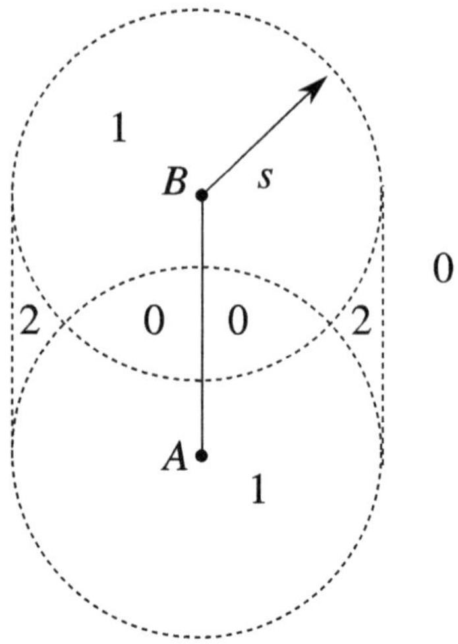

Fig. 5.12 The number of intersections of a *circle* (with fixed radius) and a line segment as a function of the *center* position

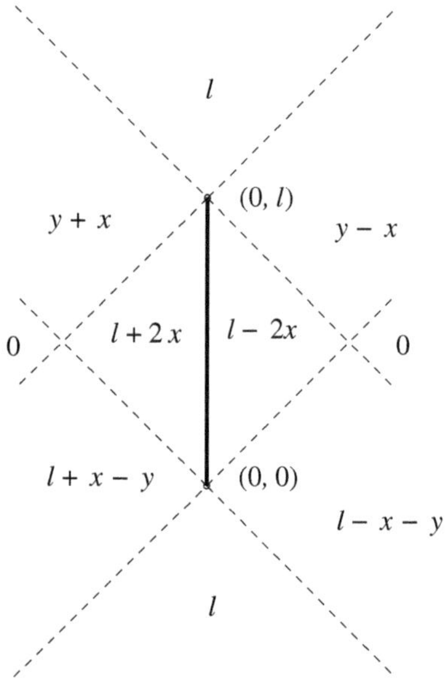

Fig. 5.13 Schematic of the square Euler-Bessel transform of a vertical line segment

Lemma 5.2 *For a compact ball A in* $\mathbb{R}^{2n}$*, the circular Euler-Bessel transform* $\mathbf{CEB}1_A$ *is a monotone increasing function on the distance to the center of A.*

Proof For x at the center of the ball A, $\chi(\partial B_r(x) \cap A) = \chi(S^{2n-1}) = 0$ for all r. □

Therefore, transforming the sum of indicator functions on several disjoint balls results in a function with local minima at the center of each ball.

For transforms of indicator functions on sets that are not balls, there is considerable variation in the Euler-Bessel transform away from the set. We call this variation

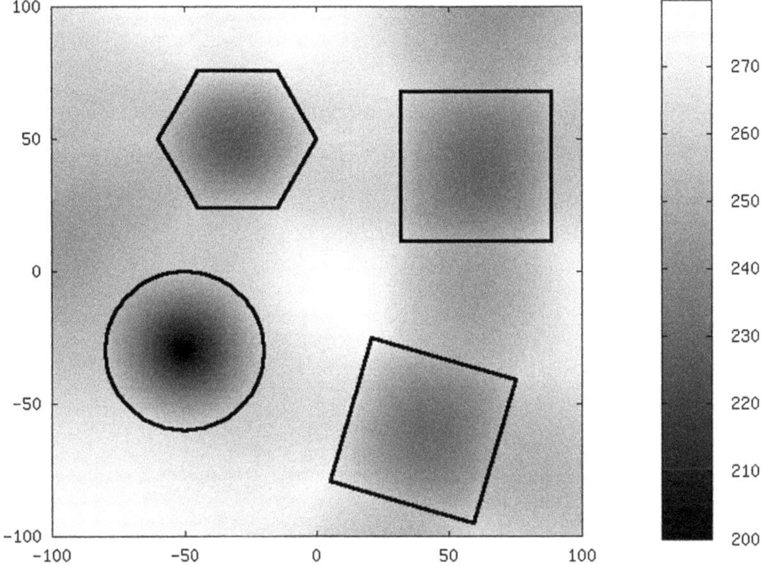

Fig. 5.14 Circular Euler-Bessel transform of indicator functions on several *regular polygons*

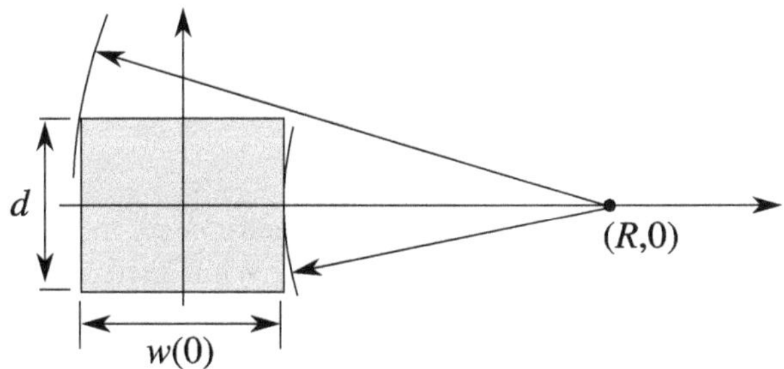

Fig. 5.15 The geometry of Proposition 5.7

sidelobe structure by analogy with the traditional Fourier transform. It happens that the sidelobe structure is entirely determined by the Euler-Fourier transform.

Proposition 5.7 *The circular Euler-Bessel transform satisfies*

$$\lim_{R\to\infty} (\mathbf{CEB}1_K)(R\cos\theta, R\sin\theta) = \mathbf{EF}1_K(\cos\theta, \sin\theta) = w(\theta). \tag{5.5}$$

Proof Without loss of generality, suppose that $\theta = 0$, and that K is contained within the rectangle $-w(0)/2 \leq x \leq w(0)/2$, $-d/2 \leq y \leq d/2$ as shown in Fig. 5.15. We therefore need to estimate the difference between the largest and smallest radii of the circles that intersect K centered at $(R, 0)$. A lower bound for this distance is

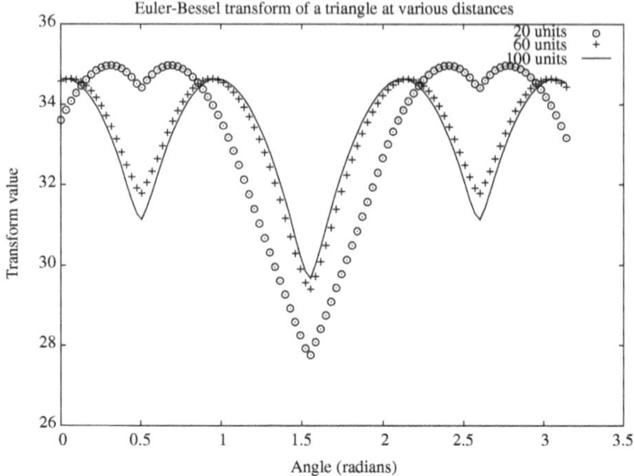

Fig. 5.16 The Circular Euler-Bessel transform of an equilaterial *triangle* with radius 10 units taken along *circles* of radius 20, 60, and 100

$$\sqrt{\left(R+\frac{w(0)}{2}\right)^2+\frac{d^2}{4}}-(R-\frac{w(0)}{2}) = R\sqrt{1+\frac{w(0)}{R}+\frac{w^2(0)}{4R^2}+\frac{d^2}{4R^2}}-R+\frac{w(0)}{2}$$
$$\sim R+\frac{w(0)}{2}+\frac{w^2(0)}{8R}+\frac{d^2}{8R}-R+\frac{w(0)}{2}$$
$$\sim w(0)$$

as $R \to \infty$. On the other hand, since K is compact and convex, the typical intersection of K with $\partial B_R(x)$ will have Euler characteristic 1. While there may be cases where this does not hold, especially at the larger radii, these will occupy zero Lebesgue measure as $R \to \infty$. □

Example 5.11 Consider the indicator function on an equilateral triangle with a radius of 10 units. The circular Euler-Bessel transform of this indicator function is shown in Fig. 5.16. Near the triangle, its Euler-Bessel transform reflects the curvature of the circular contours, though farther away from the triangle the values of the transform approach the Euler-Fourier transform. See Fig. 5.9 for comparison with the Euler-Fourier transform.

5.3.3 Sidelobe Cancellation

The Euler-Bessel transform of the indicator function on a convex set will in general not be as simple as that of a disk. Although within an inscribed circle, Lemma 5.2 ensures that the Euler-Bessel transform will have a local minimum, it may have substantial variation outside. By analogy with Fourier signal processing, we call

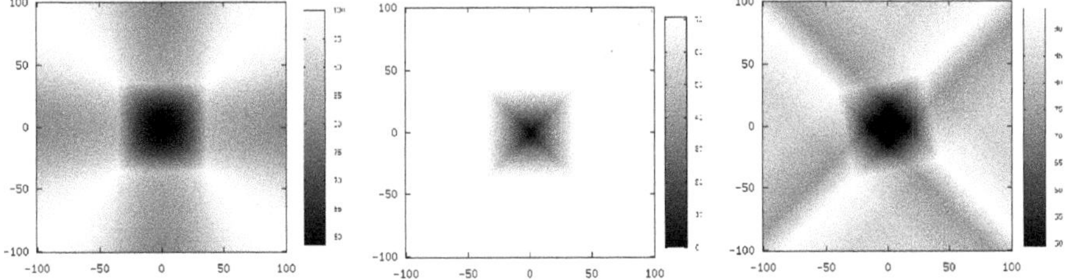

Fig. 5.17 Euler-Bessel transform of an indicator function of a *square*; *circular isoclines* (*left*), *square isoclines* (*middle*), *rotated square* with *square isoclines* (*right*)

variation of the Euler-Bessel transform of a function h outside the support of h *sidelobe structure*. In this way, the Euler-Fourier transform describes *far field sidelobe structure* of the Euler-Bessel transform.

Sidelobes are readily apparent in the left frame of Fig. 5.17, in which a local maximum for the far field occurs adjacent to each vertex of the square. This is a general fact; considering the formula for the Euler-Bessel transform of a line in (5.4), it is clear that each endpoint results in a higher-than-average contribution to the Euler-Bessel transform.

Like the matched filtering in the Fourier transform, one can adjust the integral transform to cancel the sidelobes. Briefly, the idea is that *matched filtering for Euler integral transforms amounts to choosing isoclines that are level surfaces of the functions of interest.* Therefore, if we consider the square Euler-Bessel transform of the indicator function on a square, as shown in the middle frame of Fig. 5.17, the sidelobes vanish. They reappear if the square is rotated (right frame of Fig. 5.17), however, because the square Euler-Bessel transform is not rotation invariant.

A well-established method for canceling sidelobes in matched filtering is to tailor the filter over the extent of the signal. The result is a *nonlinear* transform that preserves many of the invariances of the original transform. This procedure is called *spatially variant apodization* (Stankwitz et al. 1994; Carrara et al. 1995).

Definition 5.11 Suppose that $P_s : M \times N \to \mathbb{R}$ is a family of smooth maps from manifolds M and N, parametrized by s, that $k_1 : N \times \mathbb{R} \to \mathbb{R}$ is smooth, and that $k_2(x, \cdot) \in CF(M, \mathbb{Z})$. Then the *Euler integral transform* **SVA T** *spatially variant apodized by* P_s is given by

$$\textbf{SVA T}h(x) = \inf_s \int_{-\infty}^{\infty} k_1(x, r) \int_{\{y \in M : P_s(y,x) = r\}} k_2(x, y) h(y) d\chi(y)\, dr.$$

As an obvious special case, there is the spatially variant Euler-Bessel transform.

Definition 5.12 Suppose that N_s is a family of norms on $\mathbb{R}^n$ parametrized by s. The *Euler-Bessel transform, spatially variant apodized by* N_s is given by the formula

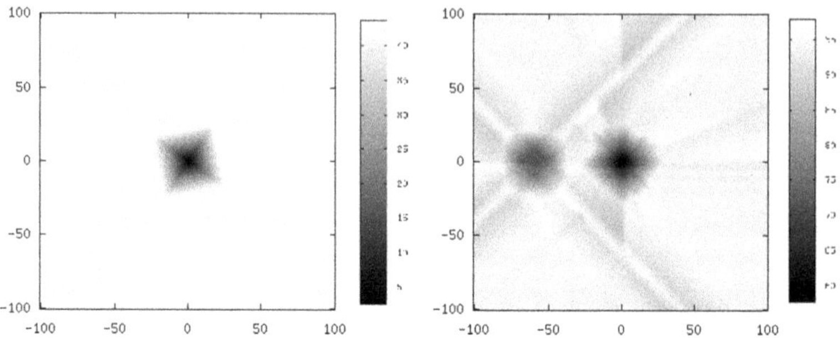

Fig. 5.18 SVA applied to *square* Euler-Bessel transform with parametrized rotation applied to a single *rotated square* (*left*) and a pair of *squares* with different rotation angles (*right*)

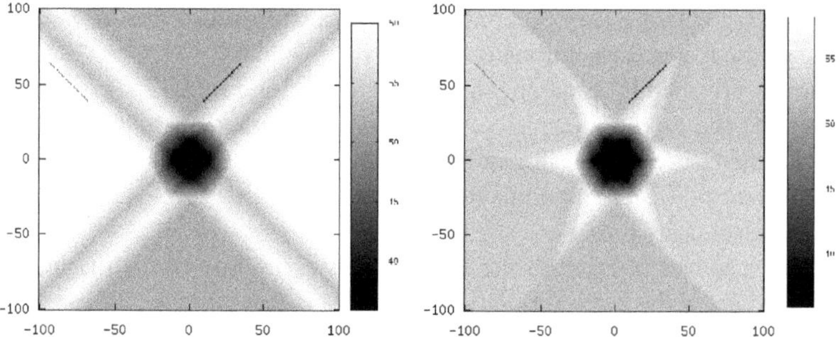

Fig. 5.19 Square Euler-Bessel transform of a *hexagon* (left) and SVA applied to square Euler-Bessel transform with parametrized rotation applied to the same *hexagon* (*right*)

$$\mathbf{SVA\ EB}h(x) = \inf_s \int_0^\infty \int_{N_s(x-y)=r} h(y)d\chi\ dr.$$

Consider the family of norms $N_\theta(x) = \|R_\theta x\|_2$ given by applying a rotation R_θ before computing the ∞-norm in $\mathbb{R}^2$. In this case, the square Euler-Bessel transform can be apodized by this family of norms, to yield a rotation-invariant form of the square Euler-Bessel transform. Figure 5.18 shows the application of this transform to the indicator function on two disjoint squares. Evidently, the sidelobes of each square have been greatly reduced.

Even if the isoclines are not perfectly matched to the shape of the support, the application of **SVA EB** may dramatically reduce the sidelobes. For instance, Fig. 5.19 shows the application of this transform to the indicator function on a hexagon. The sidelobes persist close to the hexagon, but they are substantially less prominent far away from it.

5.4 Case Study: Shape Recognition in Computer Vision

By selecting the families of isoclines to be used in spatially variant apodization, it is possible to develop *shape filters* using the Euler-Bessel transform. Suppose that a constructible function $h \in CF(\mathbb{R}^2, \mathbb{Z})$ is given, and is known to consist of a sum

$$h = \sum_{i=1}^{N} 1_{A_i}$$

of an unknown number N of indicator functions on compact, convex sets A_i. We do not know the shape of each A_i, but we do have a finite dictionary $\mathscr{D} = \{D_k\}$ of possibilities for each. That is, each A_i is a rotated, translated, dilated copy of an element of $\mathscr{D}$. The task is to automatically count, locate, and classify the A_i given the function h. For instance, we might be given the function that is the sum of indicator functions on a disk and a square.

The theory of Euler integral transforms can provide an incomplete answer to this problem. First, we can use the Euler integral to determine the precise value of N as described in Sect. 5.2. Suppose that for a set $B \subset \mathbb{R}^2$,

$$sB + x = \{x + sb : b \in B\}$$

is set obtained by dilating B by s and then translating it by x. If R_θ is a rotation by θ, Then for each $D_k \in \mathscr{D}$, we can define a transform

$$\mathbf{FD}_k h(x) = \inf_\theta \int_0^\infty \int_{x+sR_\theta D_k} h(y) d\chi \; ds.$$

If the A_i are disjoint, or nearly so, $\mathbf{FD}_k h$ will have local minima at the center of each. If we consider a minimum at a location x, there will be a local minimum for the each of the k at or near x also, by convexity of each A_i. The k that yields the smallest value of $\mathbf{FD}_k$ at x will indicate which D_k is likely to be centered at x.

Figure 5.20 shows two $\mathbf{FD}_k$ transforms of a function h that is the indicator function on several geometric figures. Notice that in the left frame (circular Euler-Bessel transform), the global minimum value occurs at the center of the disk. In the right frame (the spatially variant apodized square Euler-Bessel transform), the global minimum has been switched to the center of the square.

5.5 Open Questions

1. The analysis of discretization error in the Euler integral is in its infancy. Little work has extended beyond the observations of Baryshnikov and Ghrist (2009). One such example are the statistical results discussed in Krupa (2012), which

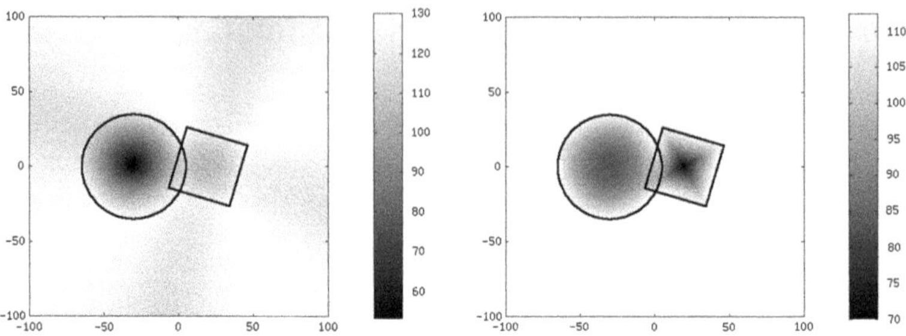

Fig. 5.20 The *circular* (*left*) and *apodized square* (*right*) Euler-Bessel transform of the indicator function on two convex sets

analyze the frequency and expected values for errors when target supports are discretized disks. The expected value for the errors allows for a bias compensation and reduction in overall error when using the Euler integral for counting targets. However, this compensation is strongly tied to the shape of the target supports, so it is unclear how to proceed if the shape of each target support is complicated or unknown.

2. Computation of the Euler-Fourier and Euler-Bessel transforms of indicator functions on piecewise linear functions is straightforward, aided by the explicit formulas. For more complicated functions, especially those that arise from level sets of images, is there a faster way to compute the transforms? It is known that the Euler-Fourier and Euler-Bessel transforms "concentrate" the Euler characteristic valuation onto a small number of points, each to be counted with a special weight. Because of this, it seems likely that Morse-theoretic reductions, such as those used in (Harker et al. 2013; Mischaikow and Nanda 2013) could simplify the required computations of the Euler characteristic substantially.
3. As a related problem, how can arbitrary $\mathbf{FD}_k$ transforms be constructed efficiently? It seems clear that explicit formulas could be constructed, but they would still seem to require an approximate Lebesgue integration step, which is slow to compute. In contrast, the explicit formulas exhibited for the circular and square Euler-Bessel transforms include an exact computation of the Lebesgue integral.
4. Because the Euler characteristic of a cell complex consisting of a single point is 1, Euler integral transforms are can be very sensitive to noise. Is there a way to mitigate this sensitivity? Two obvious possibilities exist. The first is to extend Euler integral transforms to *definable functions* using the definable Euler integral described by Baryshnikov and Ghrist (2010). Initial experiments indicate that applying a smoothing filter (such as the boxcar filter) can improve the estimates of target counts obtained using the definable Euler integral, but an explanation for why this occurs is not currently available. The second is to use the ideas of persistent sheaf cohomology developed in Chap. 6 to increase robustness.
5. Discover and apply other interesting Euler integral transforms for image processing tasks. This might be helpful in the context of automated target recognition.

References

Antimirov MY, Kolyshkin AA, Vaillancourt R (1993) Applied Integral Transforms. AMS, Springer, New York

Baryshnikov Y, Ghrist R (2009) Target enumeration via Euler characteristic integrals. SIAM J Appl Math 70(3):825–844

Baryshnikov Y, Ghrist R (2010) Euler integration for definable functions. Proc Natl Acad Sci 107(21):9525–9530

Baryshnikov Y, Ghrist R, Lipsky D (2011) Inversion of Euler integral transforms with applications to sensor data. Inverse Prob 27(12):124, 001

Bobrowski O, Borman MS (2012) Euler integration of gaussian random fields and persistent homology. J Topol Anal 4(1):49–70

Carrara W, Goodman R, Majewski R (1995) Spotlight synthetic aperture radar: signal processing algorithms. Artech House, Norwood

Chen B (2004) A simplified elementary proof of Hadwiger's volume theorem. Geom Dedic 105(1):107–120

Curry J, Ghrist R, Robinson M (2012) Euler calculus and its applications to signals and sensing. In: Zomorodian A (ed) Proceedings of symposia in applied mathematics: advances in applied and computational topology

Dimca A (2004) Sheaves in topology. Springer, New York

Ghrist R, Robinson M (2011) Euler-Bessel and Euler-Fourier transforms. Inverse Prob 27(12):124,006

Harker S, Mischaikow K, Mrozek M, Nanda V (2013) Discrete Morse theoretic algorithms for computing homology of complexes and maps. to appear in, Foundations of Computational Mathematics

Hatcher A (2002) Algebraic topology. Cambridge University Press, Cambridge

Kashiwara M, Schapira P (1990) Sheaves on manifolds. Springer, Berlin

Krupa S (2012) Numerical analysis of target enumeration via Euler characteristic integrals: 2 dimensional disk supports. arXiv, preprint arXiv:1202.2934

Mischaikow K, Nanda V (2013) Morse theory for filtrations and efficient computation of persistent homology. Discrete Comput Geom

Schapira P (1995) Tomography of constructible functions. In: 11th international symposium on applied algebra, algebraic algorithms and error-correcting codes, pp 427–435

Schürmann J (2003) Topology of singular spaces and constructible sheaves, vol 63. Birkhauser, Basel

Shepard A (1980) A cellular description of the derived category of a stratified space. Ph.D. thesis, Brown University

Stankwitz H, Dallaire R, Fienup J (1994) Spatially variant apodization for sidelobe control in SAR imagery. In: IEEE national radar conference, pp 132–137

Van Den Dries L (1998) Tame topology and o-minimal structures, vol 248. Cambridge University Press, Cambridge

Chapter 6
Noise

This chapter will

1. Develop the theory of *persistence* for cohomology to improve its robustness as a detector,
2. Demonstrate that the resulting detector can identify topological features such as acoustically opaque obstacles or quasiperiodic signals in laboratory data,
3. Exhibit theoretical bounds on the detector's robustness in terms of its required sampling density, and
4. Show that these theoretical bounds are necessary to guarantee good detection performance on experimental data.

All experimental measurements contain errors, so error tolerance is a desirable feature of any signal processing system. At first sight, topological methods appear to be both very tolerant and very intolerant to errors. This tension is characterized by the fact that unknown deformations in the configuration of sensors can produce little or no effect on the output of a topological filter or detector, yet deformations in the values returned by each sensor can dramatically change the output of both topological filters and detectors.

A deformation in the base space of a sheaf of local signals usually does not change its cohomology, via the Vietoris mapping theorem (Theorem 4.2). On the other hand, errors in the values of local sections can prevent them from extending to global sections. As a result, cohomology as defined in Chap. 4 can be extremely sensitive to noise. For instance, the number of connected components of the set of points where a function exceeds a threshold is very sensitive to noise as shown in Fig. 6.1. Since the cohomology of the constant sheaf over a space detects the number of connected components, it is sensitive to this kind of effect.

We can improve the error tolerance of topological detectors (cohomology in particular) by changing how the measurement values are interpreted. Rather than taking a statistical approach, which assumes a model of randomness associated to each measurement and its relation to those nearby, we assume that whatever values are taken must be self-consistent. In essence, we are not interested in the particular distribution of values, but rather the possible deformations of the true values. From a traditional

M. Robinson, *Topological Signal Processing*,
Mathematical Engineering, DOI: 10.1007/978-3-642-36104-3_6,

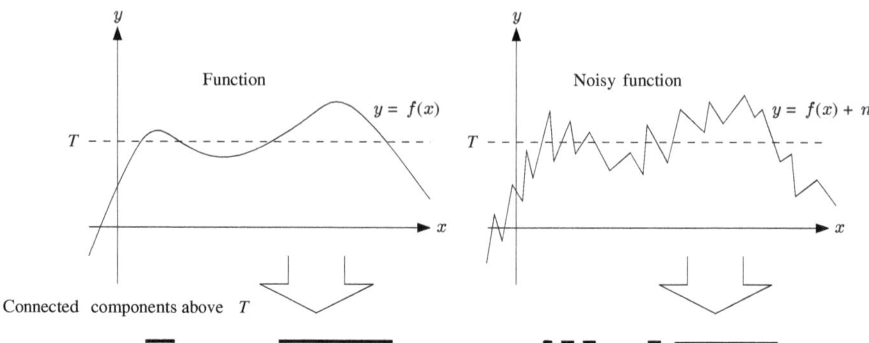

Fig. 6.1 The set of points where a non-noisy function exceeds a threshold has two connected components (*left*). The same function with noise yields a level set with many more connected components (*right*)

perspective, topological treatment of noise appears to address uniform distributions of measurements only. This is a misunderstanding, however; while the measurements themselves will be assumed to have sharp bounds on their values, their relationships to one another will be substantially more flexible.

In this way, we will introduce geometry into the signal models under discussion. The sharp bounds on each measurement taken will essentially amount to a signal-to-noise ratio. Those collections of measurements that are locally or globally consistent will be visible over large ranges of possible signal-to-noise ratio values. This gives a way to both assess data quality and the applicability of a topological filter or detector to that data simultaneously.

In most measurement systems, the resulting output is often organized into a *point cloud*, a discrete subset of $\mathbb{R}^n$. Considered as a topological space, this has exactly one topological invariant: the number of points. However, considered as a metric space—with distances between points—there is more useful topological structure available. For example, although both point clouds in Fig. 6.2 have the same number of points, the one on the left appears to describe a circle, while the one on the right appears to suggest a disk.

In this chapter, we make these appearances precise. *Persistent cohomology* ascribes a sheaf to a point cloud that encapsulates its geometry, and therefore topology. This sheaf is a detector in the language of Chap. 4, but it is substantially more robust than cohomology alone. A convenient visual representation of persistent cohomology allows for quick, quantitative interpretation of this sheaf, and the identification of significant topological features of the point cloud. Supposing that the point cloud is a discrete subset of a cell complex or manifold, persistent homology can provide precise guarantees about the homology or cohomology of this underlying space.

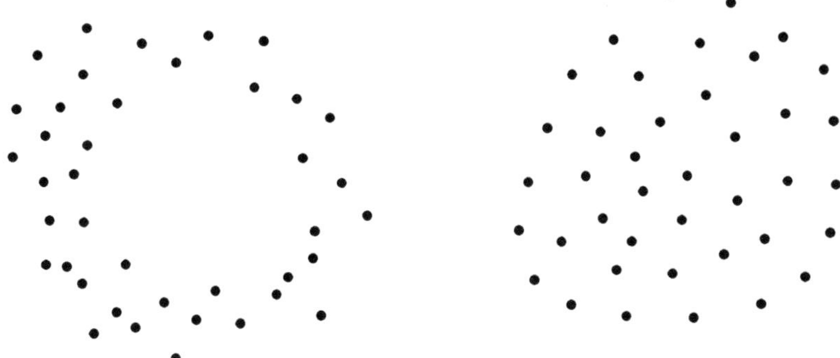

Fig. 6.2 Two finite subsets of the plane with the same number of points: (*left*) one that appears to be subsampling of a *circle*; (*right*) one that appears to be a sampling of a *disk*

6.1 Persistence

Persistent cohomology provides some measure of robustness against distortions in the stalks of a sheaf by segregating those features that "persist" for a large range of perturbations from those that are present only for a small collection of perturbations.

6.1.1 Persistence Sheaves

Suppose that the following is a sequence of topological filters, one after another

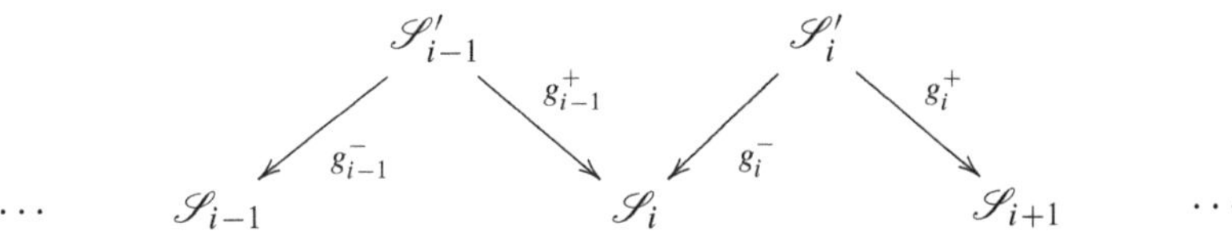

Notice that the bottom row consists of the input and output of each filter and the top row contains the state of each filter. This induces a similar looking sequence

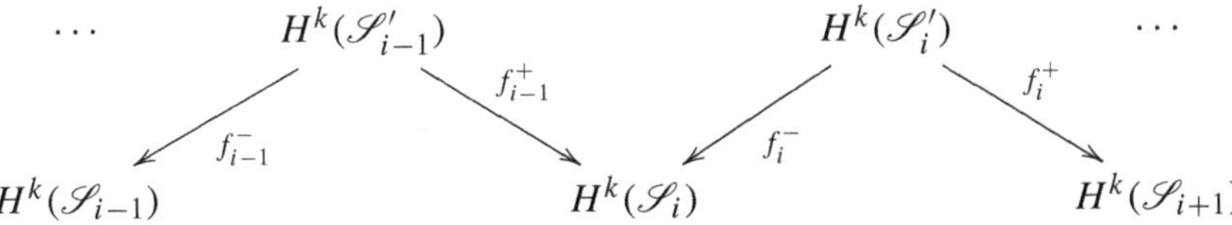

of linear maps between their cohomologies. It is immediate that this diagram itself is a sheaf $\mathscr{P}\mathscr{S}^k$, called a *persistence sheaf* over $\mathbb{R}$. For this sheaf, the base space has the cellular decomposition of $\mathbb{R}$ in which the vertices are $\mathbb{Z}$ and open intervals of the form $(n, n+1)$ are the edges. In this case, the stalks over the vertices are the spaces $H^k(\mathscr{S}_i')$, the stalks over the edges are the spaces $H^k(\mathscr{S}_i)$, and the restrictions $f_i^{\pm}$ are the maps induced by the sheaf morphisms $g_i^{\pm}$.

Proposition 6.1 *The persistence sheaf is a detector: a functor from a category of sequences of topological filters to the category of sheaves over* $\mathbb{R}$.

Proof A morphism in the category of sequences of topological filters is a diagram

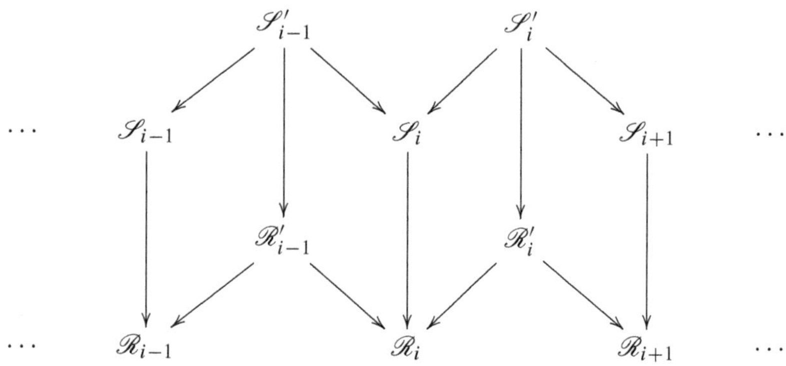

of sheaf morphisms in which all of the squares commute. This diagram induces a similar commutative diagram on the level of the cohomologies of the sheaves. This immediately satisfies the definition of a morphism of sheaves on $\mathbb{R}$. □

Sections of the persistence sheaf $\mathscr{P}\mathscr{S}^k$ have a useful interpretation. Suppose that c is a nonzero element of $H^k(\mathscr{S}_i')$. Two questions are pertinent to the significance of this element in the overall sequence:

1. Is $f_i^- c \in H^k(\mathscr{S}_i)$ in the image of f_{i-1}^+? If not, then we say that c is *born* at time i in the sequence.
2. Is $f_i^+ c = 0$ in $H^k(\mathscr{S}_{i+1})$? If so, we say that c *dies* at time $i+1$.

The *age* of c is the difference between its death and birth times. The principle of persistence is that *elements with longer ages represent more significant features*.

If all of the $\mathscr{S}_i$ and $\mathscr{S}_i'$ are sheaves of vector spaces, then the diagram of their cohomologies above is called a *zig-zag*. It then satisfies the following interval decomposition theorem, which essentially states that interpreting sections of $\mathscr{P}\mathscr{S}^k$ in terms of birth and death is canonical.

Theorem 6.1 *(Interval decomposition, (Carlsson, de Silva 2010, Derksen and Weyman 2005, Gabriel 1972)) Every persistence sheaf $\mathscr{P}\mathscr{S}^k$ arising from a sequence of sheaves of $\mathbb{F}$-vector spaces can be written as the sum (see Definition 3.8) of sheaves called* intervals, *which are $\mathbb{F}$-sampling sheaves (see Example 3.3) supported on a connected collection of vertices and edges. This decomposition is canonical; any two such decompositions are related by a permutation of their summands.*

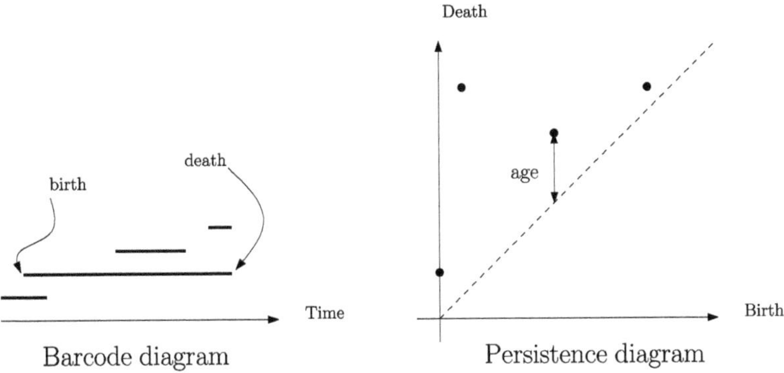

Fig. 6.3 A barcode diagram (*left*) and a persistence diagram (*right*) for a persistence sheaf with four intervals

The interested reader should consult (Carlsson, de Silva 2010) for the proof of Theorem 6.1, which is a self-contained account.

Remark 6.1 The interval decomposition fails (de Silva et al. 2011) to be canonical if the sheaves $\mathscr{S}_i$ and $\mathscr{S}_i'$ are not sheaves of vector spaces.

Two graphical representations have become popular for representing the interval decomposition of a persistence sheaf: *barcode* and *persistence diagrams*. They both permit rapid, intuitive assessment of the significant (longer aged) features and the prevalence of less significant features.

For a given cohomology degree k, the *k-barcode diagram* consists of a multiset[1] of horizontal "bars," one for each interval in the decomposition of $\mathscr{P}\mathscr{S}^k$. Each bar starts at its birth time and ends at its death time, and therefore has its age as its length. One can visually identify significant features by locating the longest bars. The left frame of Fig. 6.3 shows a barcode diagram in which the persistence sheaf has four intervals; two more significant and two less so.

The *persistence diagram for degree k* (shown in the right frame of 6.3) captures similar information to the barcode, and is a multiset whose points are contained in

$$\{(x, y) \in \mathbb{R}^2 : x \leq y\}.$$

Each point in the persistence diagram corresponds to one of the intervals in the decomposition of $\mathscr{P}\mathscr{S}^k$ with its birth time as the first coordinate and death time as its second. The age of each interval is then easily measured as the distance from the point to the diagonal; points far from the diagonal represent more significant features.

[1] A *multiset* is a set that can contain duplicate elements.

6.1.2 Interpretation of Persistent Cohomology

Unfortunately, little is known about how to interpret the sheaf $\mathscr{P}\mathscr{S}^k$ for a general sequence of topological filters. We will therefore constrain our attention to constant sheaves of vector spaces. These have a strong connection to the topological analysis of data. In this setting, it is usually sufficient to consider a sequence of cellular maps between cell complexes

$$\cdots \longleftarrow X_{i-1} \longleftarrow X_i \longleftarrow X_{i+1} \longleftarrow \cdots$$

and constant sheaves on them with a field $\mathbb{F}$ as their stalks. This sequence of sheaves induces a sequence of sheaf cohomologies

$$\cdots \longrightarrow H^k(X_{i-1};\mathbb{F}) \longrightarrow H^k(X_i;\mathbb{F}) \longrightarrow H^k(X_{i+1};\mathbb{F}) \longrightarrow \cdots$$

Example 6.1 Consider a sequence of cellular maps below

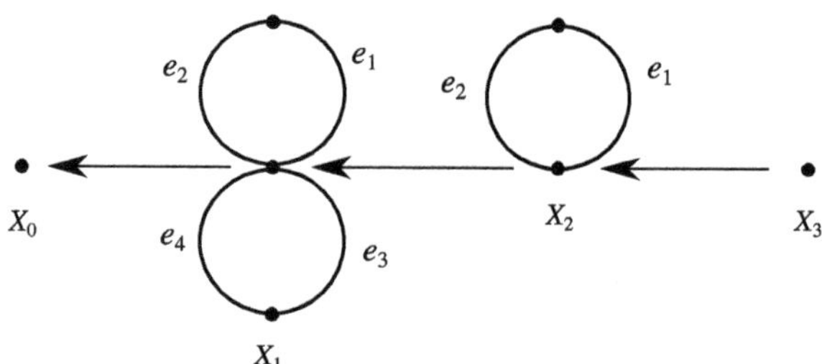

in which e_1, e_2 in X_2 maps to e_1, e_2 in X_1 (respectively). With the constant sheaf $\mathbb{R}$ over each of the cell complexes X_i, we have that

- $H^0(X_i;\mathbb{R}) \cong \mathbb{R}$. (This indicates that each space has one path connected component.)
- $H^1(X_0;\mathbb{R}) = H^1(X_3;\mathbb{R}) = 0$ since these two spaces have no edges.
- $H^1(X_2;\mathbb{R}) = \mathbb{R}$ corresponding to a value on e_1
- $H^1(X_1;\mathbb{R}) = \mathbb{R}^2$ corresponding to a value on each of e_1 and e_3.

The maps between these cohomology spaces are given by

- the identity map between each H^0 space,
- the zero map for $H^1(X_2;\mathbb{R}) \to H^1(X_3;\mathbb{R})$ and for $H^1(X_0;\mathbb{R}) \to H^1(X_1;\mathbb{R})$,
- the rank 1 map given by the matrix $\begin{pmatrix}1 & 0\end{pmatrix}$ taking e_1 to e_1, and e_3 to zero for $H^1(X_1;\mathbb{R}) \to H^1(X_2;\mathbb{R})$.

These pieces of information are summarized by the barcode and persistence diagrams in Fig. 6.4. The most significant H^0 feature is the only one; the single connected component. The most significant H^1 feature, corresponding to the longest bar in the barcode, is the loop corresponding to e_1, e_2.

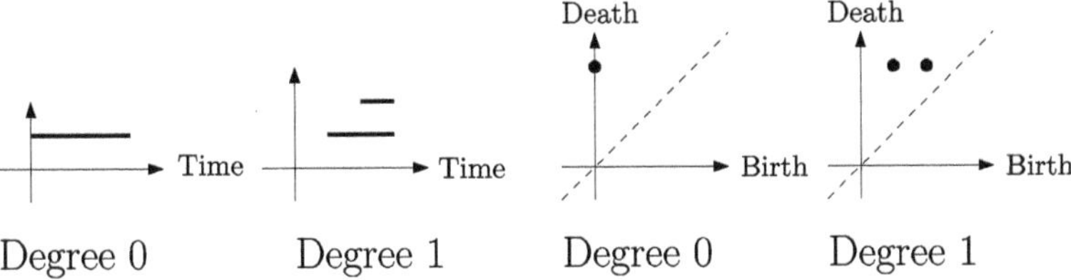

Fig. 6.4 A barcode diagram (*left*) and a persistence diagram (*right*) for Example 6.1

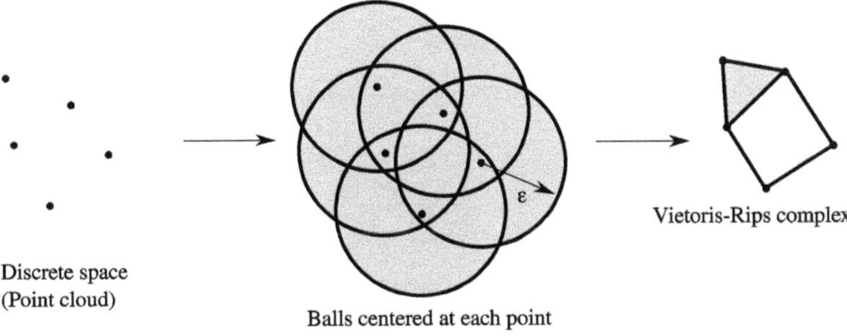

Fig. 6.5 The generation of a Vietoris-Rips complex from a point cloud

6.2 Case Study: Experimental Validation of Topology Extraction

Suppose a robot is moving in a region where it does not have access to a global position reference. This often occurs indoors or underground, since GPS satellite visibility is limited in those cases. If enough other signals are received, and their signal strength is recorded, then by the signal embedding theorem (Theorem 2.2), the collection of received signals can be used as a proxy for location. As we saw in Algorithm 2 of Sect. 4.7, this can be used to reconstruct a simplicial model of the region. We can instead use the collection of all signal measurements to produce a point cloud. The persistent homology of this point cloud can be used to identify topological features in the region, and can provide an additional layer of robustness over the simplicial model.

When interpreting point cloud data, an effective tool is the *Vietoris-Rips complex* (Fig. 6.5), which is similar to the nerve (Definition 4.17).

Definition 6.1 Suppose that P is a discrete metric space; a collection of points and a metric $d = d(x, y)$ that computes the distance between any two points. The *Vietoris-Rips complex* $\mathbf{R}_\varepsilon P$ *of size* ε is the abstract simplicial complex on P consisting of all subsets $A \subseteq P$ in which pairwise distances between points in A are less than ε.

Observe that if $\varepsilon_1 < \varepsilon_2$, then $\mathbf{R}_{\varepsilon_1} P \subseteq \mathbf{R}_{\varepsilon_2} P$. Therefore, for any increasing sequence $0 = \varepsilon_0 < \varepsilon_1 < \cdots < \varepsilon_n$, there is a sequence of cellular inclusion maps

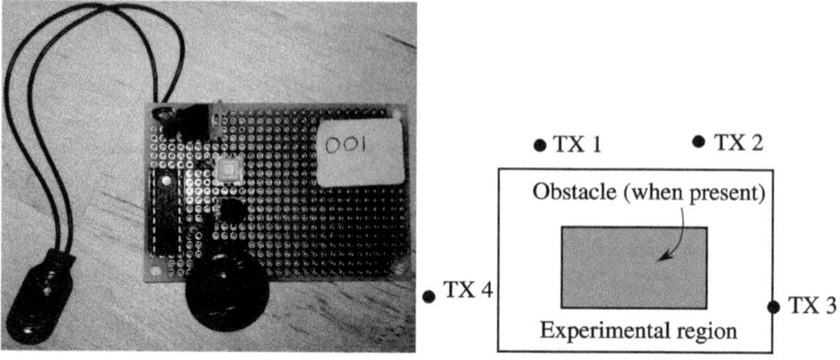

Fig. 6.6 Example acoustic transmitter (*left*) and experimental setup (*right*)

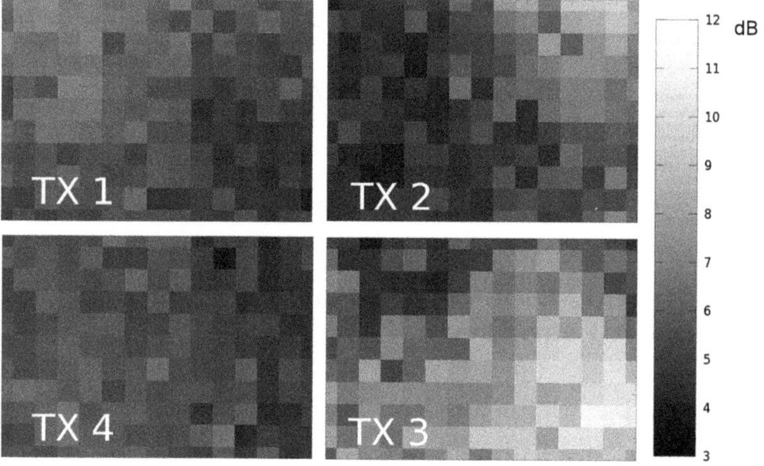

Fig. 6.7 Signal strength data collected with no obstacle

$$P = \mathbf{R}_{\varepsilon_0} P \longrightarrow \mathbf{R}_{\varepsilon_1} P \longrightarrow \cdots \longrightarrow \mathbf{R}_{\varepsilon_n} P.$$

It would seem that for a bounded set P with radius r, one should choose N evenly spaced points, namely $\varepsilon_i = r\frac{i}{N}$. Then with larger and larger N, better representations of the persistence sheaf would be obtained. However, if P is a finite set of points, there are only finitely many *critical values* for ε where the topology of $\mathbf{R}_\varepsilon P$ changes. Therefore, it is only necessary to consider a finite collection of sizes to obtain the complete representation of the persistence sheaf.

Consider the following controlled experiment, which was run in the author's laboratory (Robinson and Ghrist 2012). Four acoustic transmitters (left frame of Fig. 6.6) were placed at corners of the experimental region, as shown in the right frame of Fig. 6.6. The acoustic signal strength at a dense rectangular grid of points (3 inches between samples) in the region was measured for each transmitter. The data collected in this way is shown in Fig. 6.7; each measurement is a pixel in the grid.

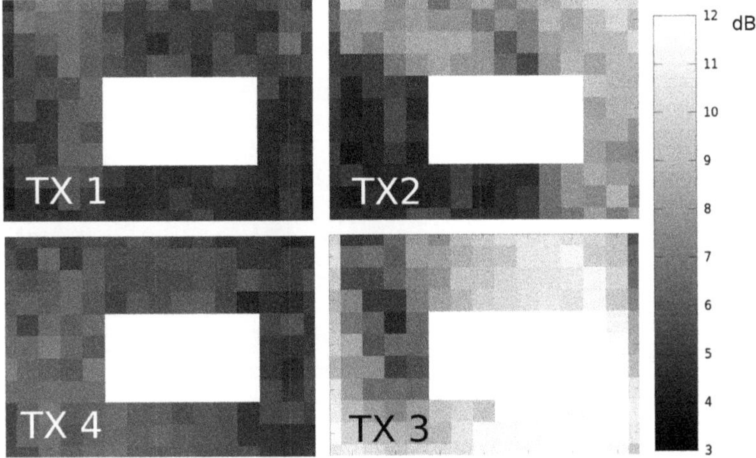

Fig. 6.8 Signal strength data collected with an obstacle. The battery of TX 3 was replaced before its collection, which caused its absolute signal strength to differ from the other collections

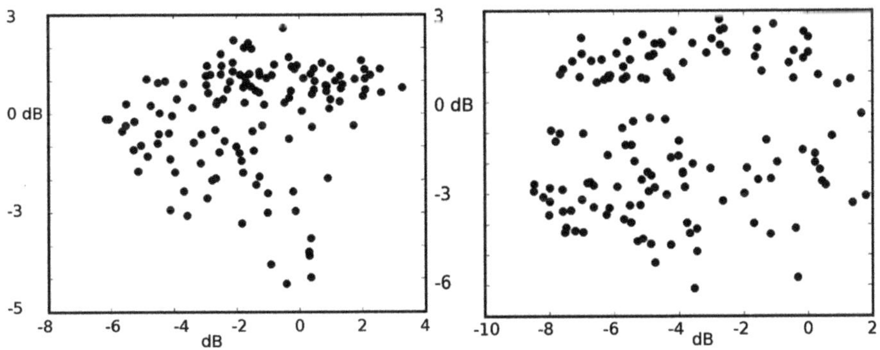

Fig. 6.9 Signal strength point cloud projected to $\mathbb{R}^2$; no obstacle (*left*), obstacle present (*right*)

To introduce a topological feature into the region, an acoustically opaque obstacle (a stack of books) was placed in the center of the region. The same collection procedure was used to measure the signal strength from each transmitter, though no data was collected inside the obstacle. This resulted in the data shown in Fig. 6.8.

Therefore, at the conclusion of the data collections, each experiment consisted of a collection of 4-dimensional vectors of signal strengths (each component represents a transmitter), one for each measurement. Therefore, we have two point clouds embedded in $\mathbb{R}^4$; can we tell which one has the obstacle? An easy, but inconclusive, way is to project both datasets to a common 2-dimensional coordinate system. The resulting point clouds are shown in Fig. 6.9.

Persistent cohomology can be computed for these point sets (using all four dimensions). For instance, the output of JPlex (Sexton and Vejdemo-Johansson 2011)

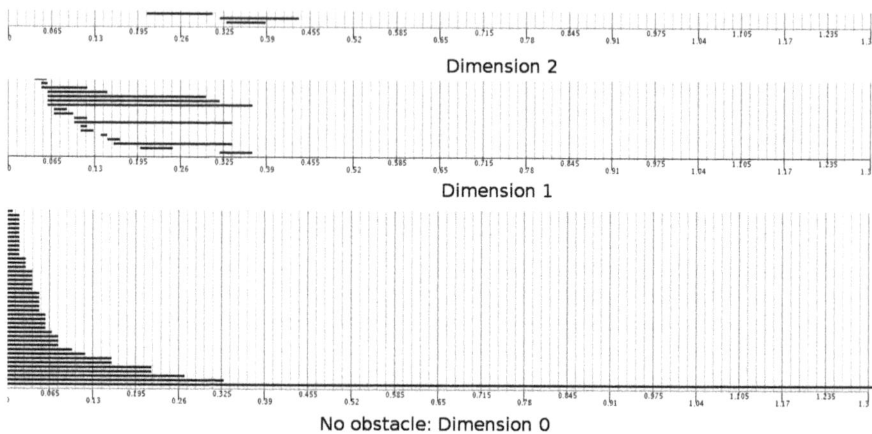

Fig. 6.10 Persistent homology barcodes for the acoustic signal strength data with no obstacle

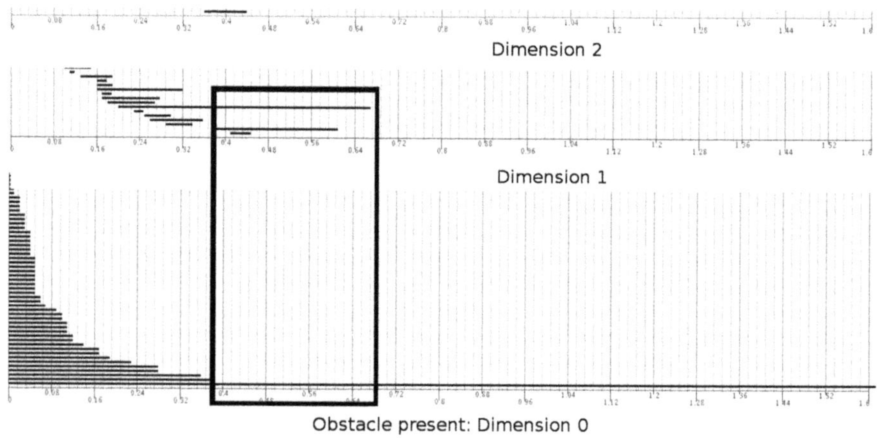

Fig. 6.11 Persistent homology barcodes for the acoustic signal strength data with obstacle present; significant 1-dimensional features are marked

is shown in Figs. 6.10 and 6.11. (The output of JPlex is a barcode diagram for persistence *homology* rather than cohomology due to relatively minor technical differences in how the computation is performed.)

Notice that even though the experimental region does not have the homotopy type of a fully 2-dimensional space, there are a few features of dimension 2. However, their ages are comparatively short, so they are not significant. On the other hand, once the short-aged 0-dimensional features vanish, there are two persistent 1-dimensional features in Fig. 6.11 that are not present in Fig. 6.10. These figures indicate the presence of an obstacle in the region, and so persistent homology is a robust detector, even in this fairly noisy setting.

6.3 Persistent Cohomology is a Robust Detector

The experimental results of Sect. 6.2 can be strengthened theoretically by a pair of results that consider continuous data, rather than sampled data. Specifically, consider the sublevel sets of a function $f : X \to \mathbb{R}$. The *sublevel set* F_a is given by $\{x \in X : f(x) \leq a\}$, so that $F_a \subseteq F_b$ whenever $a \leq b$. If X is a compact manifold and f is a *Morse function*—a smooth map f for which every critical point of has a nonsingular matrix of second partial derivatives – then the homotopy type of F_a changes at a finite set of critical values $\{c_1, \ldots, c_n\} = C_f \subset \mathbb{R}$. (See Milnor (1963); Banyaga and Hurtubise (2004) for an introduction to Morse theory.) This can be generalized to many other cases of interest; when f is piecewise linear or discrete (Forman 1998). Then if we consider the sequence of spaces

$$F_{c_1} \longrightarrow F_{c_2} \longrightarrow \cdots \longrightarrow F_{c_n},$$

any larger collection of sublevel sets will yield no new information. The persistence sheaf associated to this sequence of spaces then captures the topology of the sublevel sets of f, and therefore is a topological detector. The resulting degree-k persistence diagram $D^k(f)$ of this persistence sheaf is therefore a graphical account of the size and topology of the sublevel sets of f.

Unlike merely considering the topology of these level sets, though (which the detector defined by the formula $H^k(f((-\infty, a]); \mathbb{R})$ also does) the persistence sheaf associated to f is *robust to noise*. There is a natural metric on the space of persistence diagrams called the *bottleneck metric* that is stable under perturbations of the function f.

Definition 6.2 The *bottleneck metric* between two multisets X and Y in $\mathbb{R}^2$ is defined to be

$$d_B(X, Y) = \inf_{\gamma} \sup_{x} \|x - \gamma(x)\|_\infty,$$

where $\gamma : X \to Y$ ranges over all possible bijections. We interpret each point in a multiset that has a multiplicity greater than 1 to consist of multiple distinct copies of that point.

Theorem 6.2 *(Stability of persistence diagrams (Cohen-Steiner et al. 2007)) If f and g are Morse functions, then their persistence diagrams $D^k(f)$ and $D^k(g)$ satisfy*

$$d_B(D^k(f), D^k(g)) \leq \|f - g\|_\infty,$$

where we assume that a persistence diagram always contains the entire diagonal in $\mathbb{R}^2$ (so $D^k(f)$ and $D^k(g)$ are of the same cardinality).

The bottleneck metric is easy to see directly from two persistence diagrams. Consider the functions f and g shown in Fig. 6.12. Notice that the functions differ in that g has a local minimum and maximum. These two points give rise to another

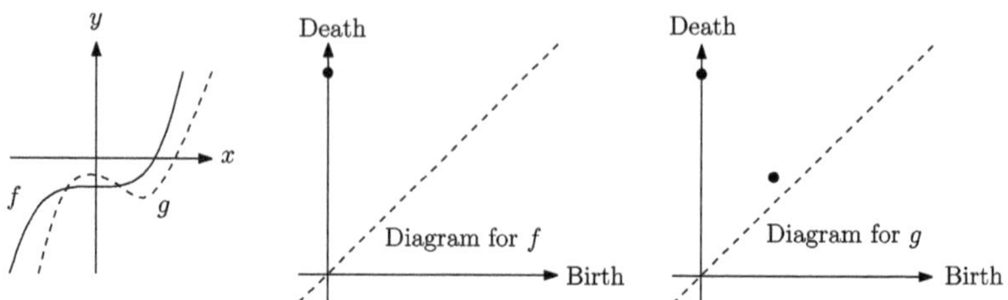

Fig. 6.12 Two similar functions and their degree-0 persistence diagrams for their sublevel sets

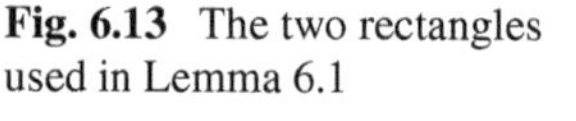

Fig. 6.13 The two rectangles used in Lemma 6.1

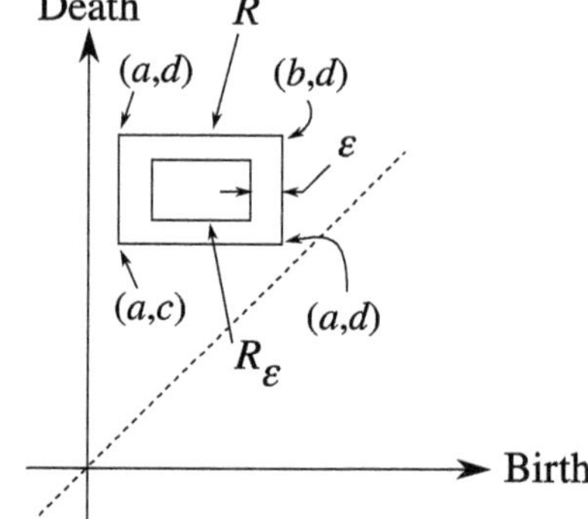

topological feature, whose age is short. This topological feature is represented by a point very close to the diagonal, so the resulting diagrams for f and g are not far apart in the bottleneck metric.

This intuition inspires the proof of Theorem 6.2 by way of a technical lemma called the Box Lemma.

Lemma 6.1 *(See Cohen-Steiner et al. (2007) for a proof) Let* $R = [a, b] \times [c, d]$ *and* $R_\varepsilon = [a+\varepsilon, b-\varepsilon] \times [c+\varepsilon, d-\varepsilon]$ *be the same box but shrunk by* 2ε *horizontally and vertically (see Fig. 6.13). If* f *and* g *are Morse functions for which* $\| f - g \|_\infty < \varepsilon$*, then the number of points of* $D^k(f)$ *that lie in* R_ε *is less than or equal to the number of points of* $D^k(g)$ *that lie in* R.

The statement of Lemma 6.1 is symmetric and f and g, so the roles of f and g may be swapped.

Proof (of Theorem 6.2) (Sketch; see Cohen-Steiner et al. (2007) for complete details) Begin by considering an easy case first, in which the bound on the bottleneck metric can be constructed explicitly. Because there are finitely many points in the persistence diagram $D^k(f)$ (not including the diagonal), we can enclose each of these points in boxes so that each box contains exactly one such point. Without loss of generality, we may assume each box is a square and each box has the same side length. Let the number $2\delta_f$ be the largest such side length for which each box contains exactly one distinct point of $D^k(f)$. This number δ_f is given by

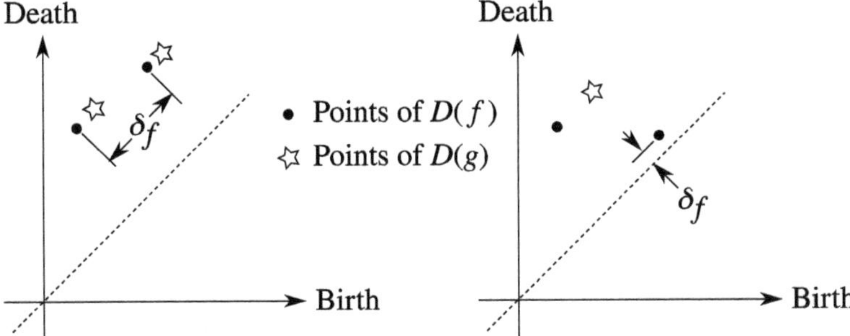

Fig. 6.14 The persistence diagram of g is *very close* to that of f in the *left* frame, but *not* very close in the *right* frame

$$\delta_f = \min\{\|p-q\|_\infty : p \in D^k(f), p \notin \Delta, q \in D^k(f), \text{ and } p \neq q\},$$

where Δ is the diagonal $\{(x, y) : x = y\}$ and $\|r\|_\infty$ is the largest of the coordinates of r.

If we consider the collection of squares centered on each point of $D^k(f)$ (now including the diagonal), each with side length $\varepsilon = \delta_f/2$, then the squares off the diagonal are disjoint from the ones centered on diagonal points and from each other.

We call a function g *very close* to f if $\|f-g\|_\infty < \varepsilon < \delta_f/2$. (See Fig. 6.14.) Let us prove the Theorem for g very close to f. In order to compute the bottleneck metric, we must construct a bijection γ between $D^k(f)$ and $D^k(g)$. Lemma 6.1 states that the number of points (counted with multiplicity) of $D^k(g)$ in each box of side length 2ε is not more than the number of points of $D^k(f)$ in the concentric box of side length 4ε. Since $4\varepsilon < 2\delta_f$, each box of side length 4ε centered on an off-diagonal point of $D^k(f)$ contains just that point (again counted with multiplicity). For each off-diagonal point p in $D^k(f)$, we let $\gamma(q) = p$ for each q in $D^k(g)$ that is contained in the box with side length 2ε centered at p.

There may be points of $D^k(g)$ that aren't covered by our definition of γ thus far, but all of these points are within a distance of ε to the diagonal. Indeed, using Lemma 6.1 the opposite way around, any point $q \in D^k(g)$ which was outside a side length 4ε box centered on every off-diagonal point in $D^k(f)$ must have *some* points of $D^k(f)$ nearby. These must be on the diagonal. Therefore, we merely make γ take these points of $D^k(g)$ to the diagonal. The resulting bijection γ moves points no farther than ε so the bottleneck metric between f and g is bounded above by ε.

To complete the proof for general f and g, the authors of Cohen-Steiner et al. (2007) reduce the Theorem to the case of piecewise linear functions, because the level sets $f^{-1}((-\infty, a])$ of piecewise linear functions have persistence diagrams with finitely many points. Since any smooth map can be approximated by a very close piecewise linear function, this suffices. The key idea is to interpolate from f to g via convex combinations, $h_t(x) = (1-t)f(x) + tg(x)$ for $t \in [0, 1]$. Because $[0, 1]$ is compact, it can be covered by finitely open intervals (a_i, b_i), for which h_{a_i} is very close to h_{b_i}. Thus, there is a sequence of very close steps from f to g, along which the Theorem holds. By the triangle inequality, this places a bound on the bottleneck metric. □

Remark 6.2 Theorem 6.2 can be strengthened if we instead consider the *p-Wasserstein distance* between two persistence diagrams $W_p(d_1, d_2)$, given by

$$W_p(d_1, d_2) = \left(\inf_{\gamma} \sum_{x \in d_1} \|x - \gamma(x)\|_{\infty}^{p} \right)^{1/p}.$$

Under this metric, the space of persistence diagrams then has the requisite structure to support statistical constructs, such as a concept of an *expected persistence diagram* (Mileyko et al. 2011).

6.3.1 Historical Context

The idea of persistent homology, much as presented here was initially presented in Edelsbrunner et al. (2002). Shortly thereafter, several sophisticated algorithms were developed for its computation, such as those described in de Silva and Carlsson (2004), Kaczynski et al. (2004), Zomorodian and Carlsson (2005), Mischaikow and Nanda (2013). The next major advance occurred with the discovery of zig-zag persistent homology by Carlsson, de Silva (2010), which was later discovered by Patel (2011) to be a sheaf. This formulation allows the use of topological filters in persistence. The use of a cohomological theory, as opposed to a homological one was pioneered by de Silva et al. (2011). Finally, we point the reader to Ghrist (2008) for a good survey article that outlines the historical details more substantially.

6.4 Case Study: Quasi-Periodic Signals

The use of topology permits the traditional rigidity of signal processing algorithms to be relaxed. Rigidity is a rather prominent feature of Fourier processing for periodic signals. For instance, suppose that f is a map $\mathbb{R} \to M$ and that for a small perturbation h of the zero function, the function g given by

$$g(t) = f(t + h(t)) \tag{6.1}$$

is periodic. We call f *quasiperiodic* in this case, and the period of g is called the *quasiperiod* of f. It is typically difficult to measure (or worse, even detect) the quasiperiod of f from its Fourier transform. These problems intensify if f is sampled, because h may not be bandlimited. The only solution to this problem would appear to be a search over possible candidates for h, which is the basis for *autofocus* techniques

(Jakowatz et al. 1996, Carrara et al. 1995). Although autofocus techniques can work in certain settings, they rely on the convergence of an inherently unstable iterative algorithm. Therefore, a more satisfactory approach is desirable.

Peristent cohomology can be used to detect and measure quasiperiods if they exist. We begin with a quasiperiodic function f from $\mathbb{R}$ to a manifold M. If M is a manifold of dimension at least 2, then the Whitney approximation theorem (Theorem 2.1) indicates that f is generically an embedding on each given quasiperiod, and an immersion if we consider its entire domain. If M is 1-dimensional, then we can use the *Takens delay map* (or quadrature signal) (Takens 1981) instead of f, namely

$$t \mapsto (f(t), f(t+\tau))$$

where $\tau \in \mathbb{R}$, to convert the image of f into a space with a loop.

Therefore, without loss of generality, let us suppose that $f : \mathbb{R} \to M$ is a quasi-periodic function that is an immersion into a manifold M of dimension at least 2. We suppose that f is sampled on some discrete set A. Because of its quasiperiodicity, then the Vietoris-Rips complex of the image $f(A)$ of this discrete set will have a significant feature in its first cohomology H^1. Therefore, quasiperiodic signals can be robustly detected.

The perturbation h in (6.1) can also be estimated from the point cloud $f(A)$ (de Silva et al. 2011). This requires a bit more sophistication, though the general idea is outlined here. A basic fact of *homotopy theory* is that the group of homotopy classes of maps[2] $M \to S^1$ is isomorphic to the first cohomology $H^1(M; \mathbb{Z})$ of the constant sheaf $\mathbb{Z}$ (Hatcher 2002). As has been discussed earlier, cohomology is not a robust detector, so the direct application of this idea is not likely to work. Inconveniently, we are also unable to apply persistent homology to the problem directly, because $\mathbb{Z}$ is not a sheaf of vector spaces so the Interval Decomposition Theorem 6.1 fails to apply.

The solution is to use the following short exact sequence of constant sheaves

$$0 \longrightarrow \mathbb{Z} \xrightarrow{\times p} \mathbb{Z} \longrightarrow \mathbb{F}_p \longrightarrow 0,$$

where p is a prime (so that $\mathbb{F}_p$ is a field). Then the resulting long exact sequence, called the Bockstein sequence, contains a homomorphism $b : H^1(M; \mathbb{Z}) \to H^1(M; \mathbb{F}_p)$, which is surjective in many cases. Therefore, we can apply persistent cohomology using $\mathbb{F}_p$ and then lift the relevant significant topological feature in $H^1(M; \mathbb{F}_p)$ to one in $H^1(M; \mathbb{Z})$. (When the desired feature is not in the image of b, the authors of de Silva et al. (2011) note that choosing a different prime p usually solves the problem.)

It remains to convert the element of $H^1(M; \mathbb{Z})$ to a map $M \to S^1$. This process requires selecting a corresponding, but smoother, element in $C^1(M; \mathbb{R})$ that is a representative of the same homology class via an optimization procedure, and then

[2] Two maps belong to the same *homotopy class* if there is a homotopy between them.

integrating this element. We refer the reader to de Silva et al. (2011) for the details of these steps.

These ideas are summarized in the following algorithmic prescription:

Algorithm 3 (Extracting circular coordinates)
Input: Values of a quasiperiodic function $f : \mathbb{R} \to M$, evaluated at $\{x_1, \ldots, x_n\}$. This is interpreted as a point cloud $A = \{f(x_1), \ldots, f(x_n)\}$ in the manifold M
Output: A map $\theta : A \to S^1$ representing the angle associated to each point in the point cloud
Procedure:

1. Select a prime p
2. While loop:
    a. Compute the persistent cohomology $H^1(A; \mathbb{F}_p)$. If this space is trivial, conclude that the function f was not actually quasiperiodic and exit with an error.
    b. Let ω be a nontrivial element of $H^1(A; \mathbb{F}_p)$.
    c. Compute the image B of the Bockstein homomorphism $b : H^1(A; \mathbb{Z}) \to H^1(A; \mathbb{F}_p)$.
    d. If $\omega \in B$, exit the loop. Otherwise select a different prime p and try again.
3. Let $w \in H^1(A; \mathbb{Z})$ be selected so that $b(w) = \omega$.
4. Replace w by a harmonic cocycle w'. (See de Silva et al. (2011) for details on this step, which results in a smoother output function θ. Skipping this step results in topologically correct output, but it is less smooth.)
5. Construct the coordinate by integating the cohomology class w' according to $\theta(x_k) = \int_{x_1}^{x_k} w'$.

6.4.1 Experimental Setup

We demonstrate Algorithm 3 by measuring the rotation rate of a ceiling fan using a simple sonar setup. The ceiling fan's speed drifted over the course of the sonar collection, but not so much as to preclude a comparison of Algorithm 3 with more traditional processing. Consider the collection geometry indicated in Fig. 6.15. In this scenario, a rotating fan was located on the ceiling of a room with walls made of drywall and a sonar platform was placed on the floor. We assume that the function that takes the angular position of the fan's blades to the train of echos received by the sonar platform is quasiperiodic, which essentially makes the simplifying assumption that the speed of sound is much faster than the fan's rotation speed. Since the sonar receiver was active during the entire collection (including during transmission), it is reasonable to represent the collection as a function $f : \mathbb{R} \to \mathbb{R}^n$ where n represents the number of time samples collected after each pulse was transmitted.

For the experiment described here, the sonar platform consisted of the speaker and microphone of a cell phone. The transmitted waveform was an impulse train with

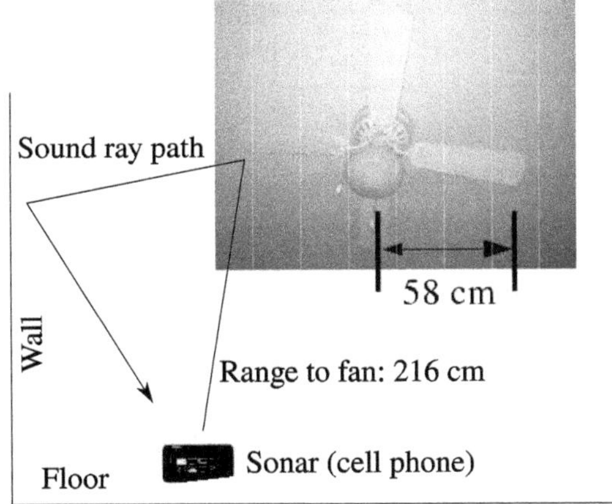

Fig. 6.15 Experimental setup: fan on ceiling, sensor on floor

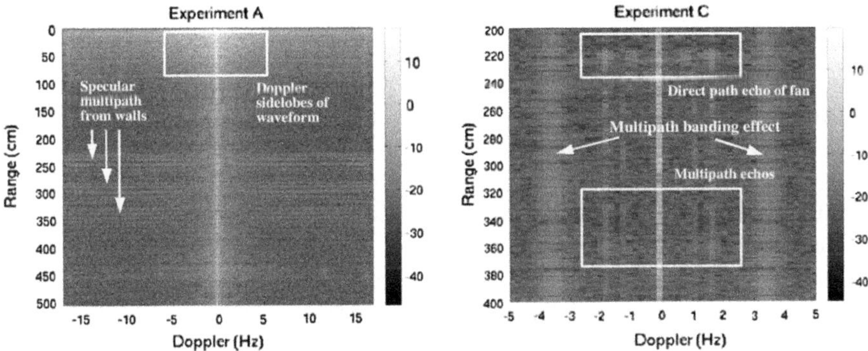

Fig. 6.16 Doppler-processed pulses: fan stopped (*left*), rotating with a rate of approximately 0.85 Hz (*right*)

a bandwidth of 7 kHz, range resolution of 5 cm, and pulse repetition rate of 34 Hz. In all, 175 pulses were collected.

Figure 6.16 summarizes the collected data, processed with traditional Fourier methods. For each pulse, the received echo train was aligned so that the earliest echo occurred at time zero. Each aligned pulse was stored as the column of a matrix R. The Fourier transform was performed along each row of R, to produce the new matrix R'. The interpretation is that each row of both R and R' correspond to a particular *range* or distance from the sonar. Each column of R' corresponds to a particular *Doppler frequency shift* due to motion of targets. The amplitude of each entry in R' corresponds to the sonar reflectivity at the corresponding range and doppler location; rigid obstacles tend to have higher sonar reflectivity. Several instances of R' are shown in Fig. 6.16, in which echos from moving fan blades are visible in the right frame as localized bright spots.

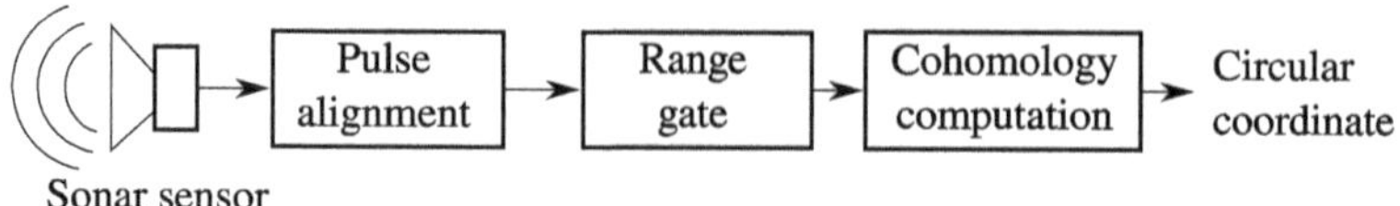

Fig. 6.17 Schematic diagram of the processing chain

Although there are several definite doppler frequencies that appear to be highlighted in the right frame of Fig. 6.16, there is some ambiguity as to the correct rotation rate. In particular, there is a marked banding effect (visible at all ranges) because the echos did not decay completely within the period of one pulse. This frustrates traditional analyses, and it is unclear if the fan rate changed over the course of the collection. Additionally, substantial blurring in doppler is present as the fan's rotation speed drifted over the course of the collection.

6.4.2 Results of Persistent Cohomology

(Thanks to Vin DeSilva for supplying the author with an implementation of Algorithm 3 to process the ceiling fan data described in this section.)

Although it is instructive to imagine processing *all* of the data, this turns out to be impractical. Most importantly, as the left frame of Fig. 6.16 shows, there is substantial receiver desensitization at small ranges since (1) the reciever was not turned off during pulse transmission and (2) there were a number of immobile, reflective obstacles close to the sonar. Because of this, at small ranges the received signal is approximately *constant*, which obscures the comparatively smaller quasiperiodic echoes from the fan blades. Because of this, the samples corresponding to small ranges were discarded; a process usually called *range gating*. However, as Fig. 6.17 indicates, no further preprocessing of the data was performed.

The output of Algorithm 3 is shown in Fig. 6.18. A persistence diagram is shown at left, in which exactly one generator of H^1 is significant (marked with a star). This corresponds to the rotation of the fan, and is confirmation of the quasiperiodic nature of its rotation.

The recovered angular coordinate is shown in the right panel of Fig. 6.18, in which a steady (if a little wobbly) progression is clear. A close examination of the this progression shows a very slight slowing of fan rotation rate as collection proceeded (later times are on the right of the diagram). Although this might be traced to the fact that the fan was spun by hand before the collection and allowed to coast during the collection, the reader is cautioned against reading too much into the slope of the angular progression. Algorithm 3 is topological in nature and the representation of the angular coordinate as a cohomology class is not unique. Indeed, the "wobbles" appearing at about 0.3 rotations are probably artifacts.

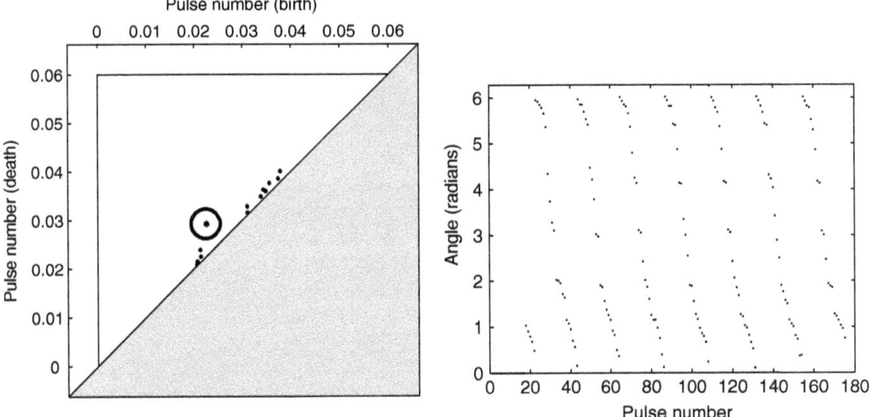

Fig. 6.18 Persistence diagram of the output (*left*) along with recovered angular coordinate (*right*). The recovered angular coordinate plot has time as its *horizontal axis* and angle in radians as its *vertical axis*. Note the significant feature in the *left panel, circled*

Usually, Fourier methods are used to detect periodic signals. However, if the signals are not periodic, Fourier methods become difficult to apply. Figure 6.19 shows the range-doppler response of a fan slowing to a stop. The resulting echos of the fan (occuring at a range of approximately 180 cm) are indistinguishable from the echos of the (static) ceiling at around 210 cm. Persistent cohomology is a highly effective solution to this problem. Figure 6.20 shows the persistence diagrams of three situations in which the fan was stopped or slowing. In the middle and right panels, the fan slowed to a stop. Observe that in the two cases where the fan was in motion, a significant generator of H^1 was present, which indicates that the echo response was quasiperiodic.

6.5 Recovering a Space from a Point Cloud

The persistent cohomology associated to a point cloud accurately represents the cohomology of its underlying space when the union of balls centered at the points in the cloud encapsulates the same topological information as the space. When this occurs, persistent cohomology is a more robust form of the sheaf cohomology defined in Chap. 4. (Otherwise, it is still a robust detector, but captures different information. For instance, the persistent cohomology of a sequence of non-constant sheaves usually does not represent the cohomology of the space.)

Stated more precisely, suppose $K \subset \mathbb{R}^n$ is a CW complex and $P \subset \mathbb{R}^n$ is a finite subset with p points. The persistent cohomology for P, with the metric $d = d(x, y)$ induced from $\mathbb{R}^n$, was defined in the Sect. 6.4. Does this detector bear any resemblance to the cohomology of a constant sheaf over K?

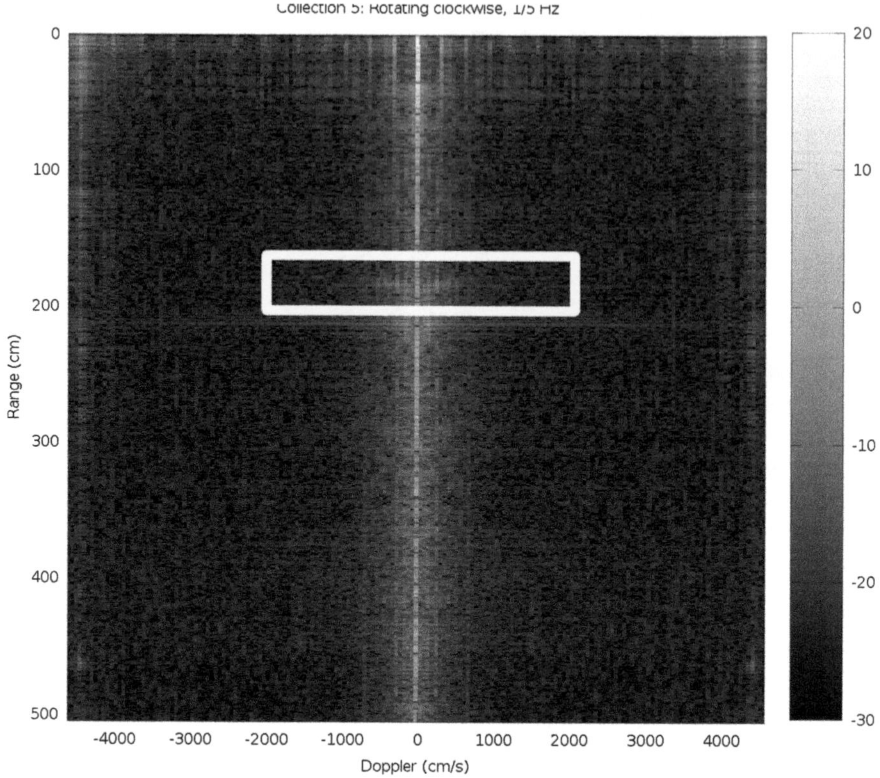

Fig. 6.19 Range-doppler plot of fan response as the fan slows to a stop. The aperiodic nature of the echos causes blurring (marked) across the doppler (*horizontal*) direction

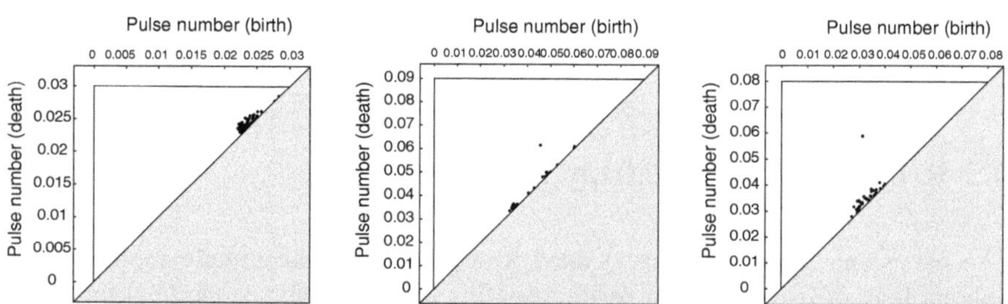

Fig. 6.20 Persistence diagrams for a fan which is stopped (*left*), spinning counterclockwise and slowing to a stop (*middle*), and spinning clockwise and slowing to a stop (*right*). Note the off-diagonal generator in both of the cases when the fan was moving

We begin by supposing that the points in P may be in error. As in the case of persistent cohomology, this error is modeled by balls of a given radius centered at each $x \in P$. The appropriate generalization of this idea is the notion of a specific kind of sublevel set, called an *offset*.

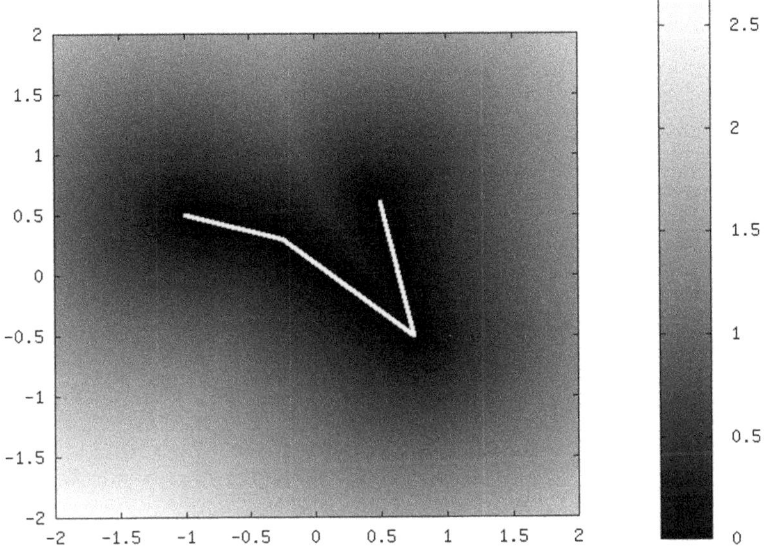

Fig. 6.21 The distance function to a collection of line segments using the Euclidean metric

Definition 6.3 Suppose that $d = d(x, y)$ is the metric on $\mathbb{R}^n$. The *distance function to* K is the function (see Fig. 6.21)

$$R(x) = \inf_{y \in K} d(x, y).$$

The ε-*offset* K^{ε} of K is the sublevel set $R_{\varepsilon} = R^{-1}([0, \varepsilon))$.

Although R is typically not a Morse function, since it may not be smooth, the robustness described in Sect. 6.4 for persistent cohomology is a good guide to the robustness properties enjoyed by offsets. The union of balls of radius ε around each $x \in P$ is P^{ε}. With this definition in hand, we can compare how geometrically different two subsets of $\mathbb{R}^n$ are.

Definition 6.4 If X, Y are subsets of $\mathbb{R}^n$, then the *Hausdorff metric* between them is

$$d_H(X, Y) = \inf_{\varepsilon}\{X \subseteq Y^{\varepsilon} \text{ and } Y \subseteq X^{\varepsilon}\}.$$

If two sets are identical except for points on their boundaries, then their Hausdorff metric is zero. If there are many sample points P, the Hausdorff metric between P and K should be small in order to be a good approximation. However, the Hausdorff metric is strictly geometric; two spaces can have small Hausdorff metric, yet be topologically very different as shown in Fig. 6.22. We therefore need some additional

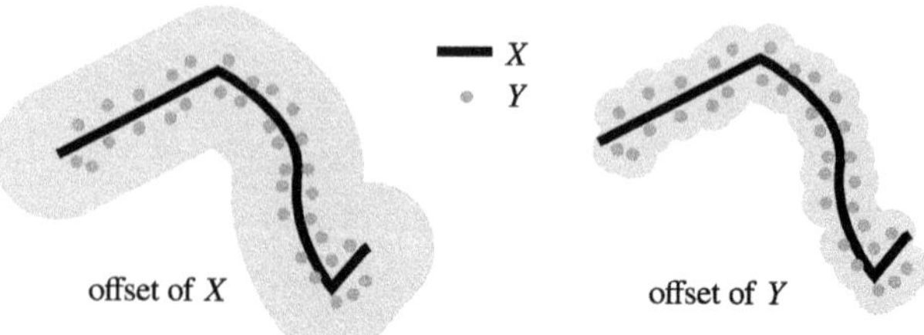

Fig. 6.22 Computing the Hausdorff metric between two topologically distinct sets

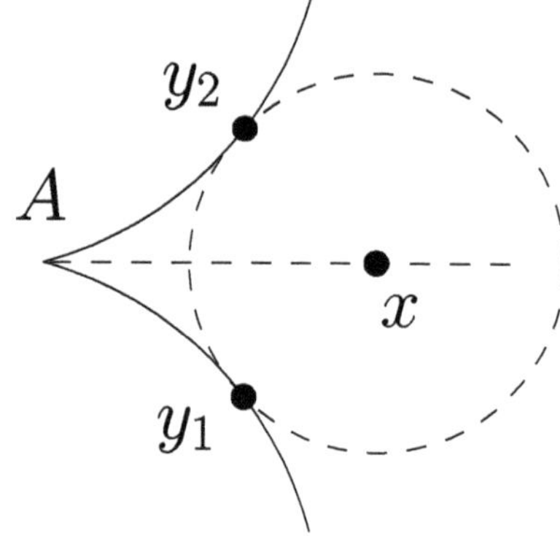

Fig. 6.23 A set A with a cusp has zero reach, because there are always multiple closest points y_1, y_2 for each point x along the bisector

geometric conditions for ensuring that the offset of P accurately reflects the topology of K. The most intuitive of these conditions is based on the concept of *reach*.

Definition 6.5 The *reach* (Federer 1959) of K is the largest μ such that if $x \in \mathbb{R}^n$ and $R(x) < \mu$, then there exists a unique element of K closest to x. We usually write $\operatorname{reach} K = \mu$.

Sets with infinite reach are convex, and sets with concave cusps such as shown in Fig. 6.23 have zero reach.

Reach is an important concept for topology because gives conditions for offsets to recover the topology of a space. Usually people work with sets whose reach is nonzero, but according to tradition we call this situation "positive reach."

Theorem 6.3 *(Tubular neighborhood theorem for positive reach (Federer 1959, Thm. 4.8)) Suppose that the reach of K is positive, then any offset K^ε is homotopy equivalent to K if $\varepsilon < \operatorname{reach} K$ as shown in Fig. 6.24.*

Proof The proof is proceeds by explicit construction; one considers a suitable generalization to the gradient vector field

$$\nabla R(x) = \frac{x - u(x)}{R(x)},$$

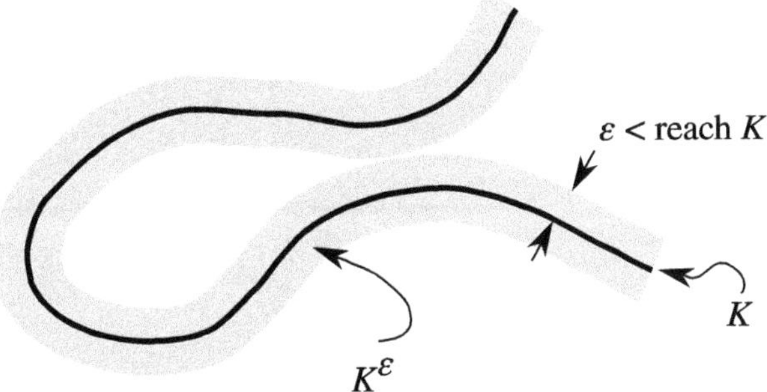

Fig. 6.24 A tubular neighborhood K^{ε} of K

where u associates x to the nearest point in K, which is unique by assumption. One merely needs to show that this vector field is smooth enough to result in a unique solution to the flow along it, and that it generates a family of diffeomorphisms. The proofs of these claims are elementary, though a little tedious, and follow from the existence and uniqueness theorem for solutions of ordinary differential equations.□

Remark 6.3 We note in passing that the notion of positive reach allows one to use offsets to generalize the notion of tubular neighborhoods of compact manifolds. For instance, (Lee 2003, Thm. 10.19) is a traditional account.

The theorem then allows us to pose a reasonable sampling density requirement: the union of balls around each P must strictly lie within K^{μ}, where μ is the reach of K. In this way, positive reach sets a bound on the required measurement accuracy. If the samples are less accurate than the reach of K, then recovery of the topology could be impossible. Similarly, if K is of finite volume, this upper bound μ on the size of each ball around points in P will set an upper bound on the volume of P^{μ} given a fixed number of points. Therefore, the minimum number of sample points can easily be determined from the volume of K and its reach.

One drawback of the tubular neighborhood theorem is that it does not explicitly treat the presence of noise in measurements. In particular, if the points in the set P do not lie on K, under what circumstances can they be used to infer topological properties of K? It is useful to quantify the error in an approximation P in terms of the reach of K.

Definition 6.6 (Chazal and Lieutier, 2006) P is an ε-approximation to K if $d_H(P, K) < \varepsilon \operatorname{reach} K$.

With this definition, we can give explicit bounds on the required sampling density.

Theorem 6.4 *(Homotopy reconstruction for submanifolds (Niyogi et al. 2004, Prop. 7.1, Chazal and Lieutier 2006, Cor. 3.3)) Suppose that K is a compact submanifold of $\mathbb{R}^n$. If $\varepsilon < 3 - 2\sqrt{2} \approx 0.17$ and P is an ε-approximation to K, then for every α satisfying*

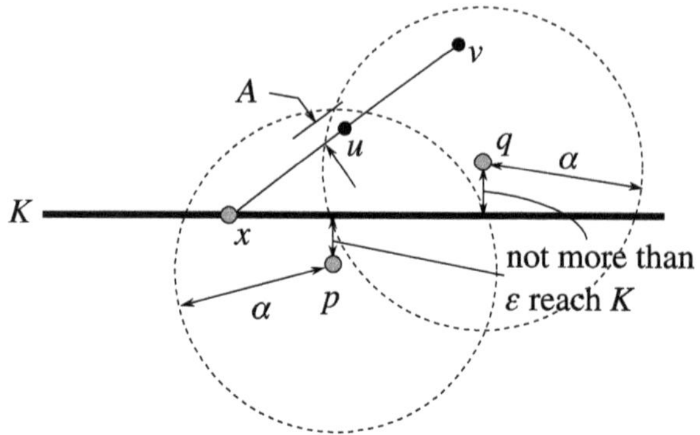

Fig. 6.25 The geometry for Theorem 6.4

$$\frac{\text{reach}K}{2}\left(1+\varepsilon-\sqrt{(1+\varepsilon)^2-8\varepsilon}\right) < \alpha < \frac{\text{reach}}{2}\left(1+\varepsilon+\sqrt{(1+\varepsilon)^2-8\varepsilon}\right)$$

the offset P^α is homotopy equivalent to the offset K^α.

Proof Suppose that $x \in K$ is given and that $v \in P^\alpha$. It suffices to show that the line segment connecting v to x lies entirely within P^α, for then v could be retracted to K. Since balls in $\mathbb{R}^n$ are convex, this statement is true whenever v lies within α of x and some point $p \in P$.

Consider the case where this does not hold, shown in Fig. 6.25. Then there must be a $q \in P$ within a distance of α to v, but farther than α from x. In order to "hand off" points along the line segment from v to x, we employ another sample point $p \in P$ that lies within a distance of α to x.

Locate the point u along the line segment connecting v and x, that is closest to the midpoint of the line segment connecting p and q. If u is inside the α-offset of $\{p, q\}$, then the entire line segment from x to p will also be so contained. We therefore place bounds on α so that u lies within P^α.

In the worst case, x, p, and v are colinear. So we must have

$$\text{reach}K - A < \alpha - \varepsilon\ \text{reach}K,$$

where $A = \sqrt{(1-\varepsilon)^2(\text{reach}K)^2 - \alpha^2}$ is the distance from u to the boundary of the α-offset of $\{p, q\}$ in the worst case. Reararanging, we find that

$$(\text{reach}K - \alpha + \varepsilon\ \text{reach}K)^2 < A^2 = (1-\varepsilon)^2(\text{reach}K)^2 - \alpha^2$$

which reduces to

$$\alpha^2 - \alpha\,\text{reach}K(1+\varepsilon) + 2\varepsilon(\text{reach}K)^2 < 0.$$

This is satisfied provided

$$\frac{\text{reach}K}{2}\left(1+\varepsilon-\sqrt{(1+\varepsilon)^2-8\varepsilon}\right)<\alpha<\frac{\text{reach}K}{2}\left(1+\varepsilon+\sqrt{(1+\varepsilon)^2-8\varepsilon}\right)$$

Of course, this requires that

$$\begin{aligned}(1+\varepsilon)^2-8\varepsilon&>0\\1+2\varepsilon+\varepsilon^2-8\varepsilon&>0\\\varepsilon^2-6\varepsilon+1&>0\end{aligned}$$

which implies that we should ensure that

$$\varepsilon<3-2\sqrt{2}.$$ □

The presence of errors ($\varepsilon>0$) impacts the sampling accuracy α required. From the bound on α in Theorem 6.4, if the noise is larger than $1/8$ of the true reach, then it is impossible to accurately recover the original manifold K. Instead, we are forced to recover a noisy estimate K'.

It is difficult (or even impossible) to estimate the reach of K since K itself is usually unknown. In practical settings one can attempt to estimate the reach using the samples P; this can suggest whether the hypotheses of Theorem 6.4 are satisfied or not. A low pass filter should be applied before making this estimate, or else the estimate of reach will likely be too small as indicated in Fig. 6.26. If too small an estimate for reach is obtained, it will suggest to sample at a much higher density than is necessary.

6.6 Case Study: Recovery of a Space from Measurements of Waves

When the hypotheses of Theorem 6.4 are satisfied, then the cohomology recovered from the point cloud is the cohomology of the underlying space. Persistent cohomology taken with a collection of radii satisfying the hypotheses will then recover exactly the same topological features; this validates the use of persistent cohomology as a proxy for cohomology. Usually, the hypotheses are difficult to verify, since the reach of the space is not known in advance, if ever. Worse, as noted by some practioners with analogous results to Theorem 6.4 (for instance Amenta et al. (2002)) the sampling hypotheses are too stringent to be met in practice. Therefore, one typically needs to rely on external sources of information to cross-validate the output of a persistent cohomology computation.

Because of these considerations, the performance of persistent cohomology in practice is best understood through example. This case study examines the recovery of the topology of a propagation environment as in Sect. 6.2, but examines the modes of failure as well. Two datasets with two different signal modalities will be examined: radio propagation and acoustic propagation, whose relevant parameters

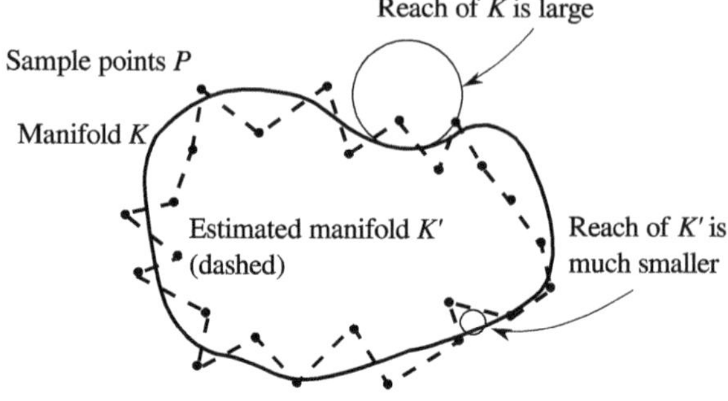

Fig. 6.26 If the underlying manifold K is unavailable, a noisy estimate K' can have much smaller reach

Table 6.1 Sampling parameters of the two scenarios in Sect. 6.6

Description	Radio scenario	Acoustic scenario
Data source	Simulation	Experiment
Frequency	2.5 GHz	4 kHz
Wavelength	12 cm	9 cm
Sample spacing	0.5 m	7.7 cm
	4.2 wavelengths	0.85 wavelengths
Noise	None simulated	Present
Ratio of reach to mean sample spacing	2.6 %	28 %

are summarized in Table 6.1. The radio propagation example is also described in Sect. 4.7, while the acoustic experiment was described earlier in this Chapter in Sect. 6.2. The wavelengths used in each experiment are similar, though the sampling densities are substantially different. The hypotheses of Theorem 6.4 are not met in either of the two scenarios, but substantially fewer samples per wavelength are used in the radio propagation scenario. This results in more reliable recovery of topological features in the acoustic scenario. In terms of the spatial sampling required, longer wavelengths can be more useful for probing the topological features of an environment.

Our experiments aim to compare the sampling density required by Theorem 6.4 with those reasonably attained in practice. Unfortunately, these requirements are stated in terms of the reach of the *unknown* underlying manifold. In order to validate performance against this criterion, we will use *a priori* knowledge of the geometry of the environment to make this determination.

The experimental procedure measures the geometry and topology of the propagation environment indirectly. As discussed in Chap. 2, the collection of received signal stengths should be thought of as a map $P : M \to S$ from the propagation environment M to a manifold of signals S. Under appropriate conditions (for instance Theorem 2.2) this map is an embedding, so the image $P(M)$ of the environment is a topologically accurate model of M. It is this image $P(M)$ that the experiments in this case study attempt to recover. Therefore, although measurements are taken by

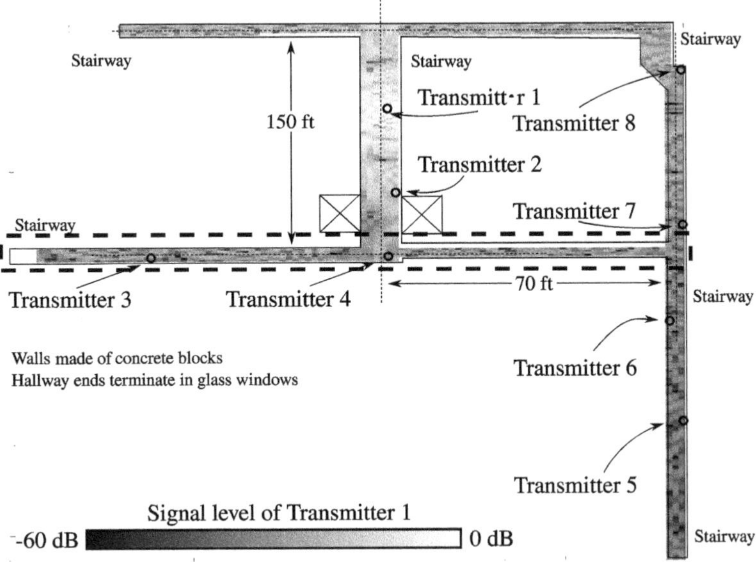

Fig. 6.27 Simulated wireless signal strength (from Transmitter 1) on the third floor of David Rittenhouse Laboratory. The hallway used to estimate manifold curvature is marked with a *dashed line*

placing a receiver at points in M, the sample density must be assessed in $P(M)$ using a metric induced from S. From a practical perspective, this can make controlling sampling density difficult.

The radio propagation scenario uses the floorplan of the third floor of the David Rittenhouse Laboratory at the University of Pennsylvania shown in Fig. 6.27. There are eight wireless network transmitters located on the walls as indicated in the figure. Using a simple model of radio propagation, the received signal strength from each transmitter was predicted to a dense grid of points spaced every 0.5 m. The signal strength for transmitter 1 is shown in grayscale in Fig. 6.27. This simulated the process of sampling the environment by moving a receiver to many locations.

Measurements along one of the hallways (marked with a dashed line in Fig. 6.27) were selected to perform analysis of the sampling density. Along this hallway, measurements were spaced 0.5 m apart in the physical model. This does not correspond directly to the distance between points in the signal manifold, however. The signal manifold distance between two points that are adjacent in physical space varies substantially, as is shown by circles in Fig. 6.28. This is the sample density α in Theorem 6.4.

Since the simulation did not include noise, we can assume that the measurements lie exactly on the signal manifold. In Theorem 6.4, we simply take $\varepsilon = 0$ and observe that sampling criterion is that the reach of the signal manifold must be larger than the signal manifold sample density. The reach of the manifold in this case is merely the least upper bound on the radii of curvature, a fact noted

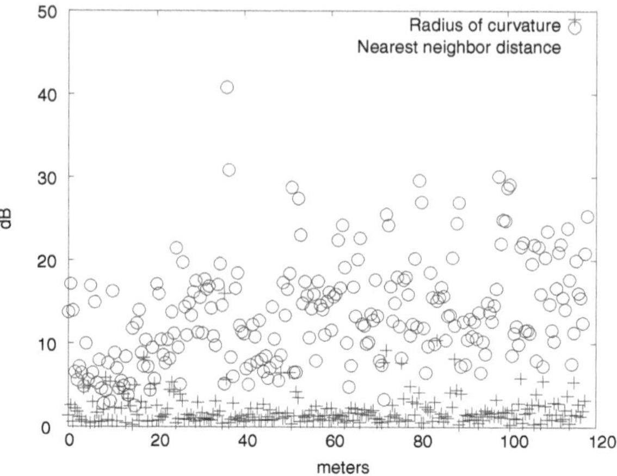

Fig. 6.28 Comparison between the distance between adjacent points (*circles*) and the local radius of curvature (*crosses*) along the hallway marked in Fig. 6.27

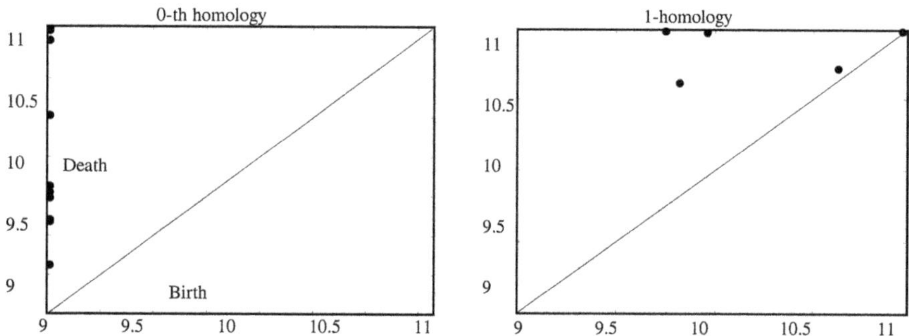

Fig. 6.29 Persistent homology diagrams of the simulated third floor of David Rittenhouse Laboratory (axes are in decibels)

in Niyogi et al. (2004, Section. 6). The local curvature of the signal manifold (at each sample point) is plotted as crosses in Fig. 6.28. Clearly the sampling criterion is not met essentially anywhere on the hallway under discussion. Indeed, the average sample density is less than 3 % of that which is required by the sampling criterion.

The fact that this scenario is so undersampled explains why in Sect. 4.7 the nerves are very sensitive to the choice of threshold. It is instructive to test persistent cohomology as well. The results of 0- and 1-degree homology are shown in Fig. 6.29. There are no significant features in the persistent 1-homology, which contradicts the fact that the hallways in Fig. 6.27 form a loop. Clearly the sampling density was too low.

In contrast, although the acoustic data presented in Sect. 6.2 are undersampled, they are *less* undersampled. Figure 6.30 shows the estimated curvature (left) and adjacent distances (right) for the dataset with no obstacle. Both plots use the same

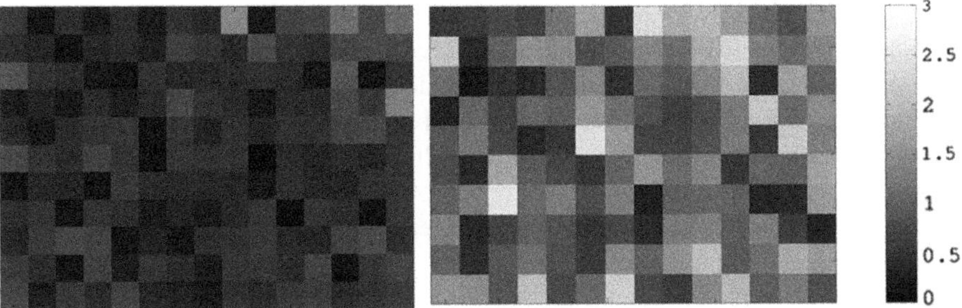

Fig. 6.30 Local estimates of curvature (*left*) and nearest neighbor distance (*right*) both color scales are in decibels

Fig. 6.31 Two sets of points near the surface of a narrow cylinder; is the hollow center significant or not?

colorscale for consistency. As Table 6.1 shows, the sample density is about 28 % of what is required. On the other hand, since this dataset contains noise, the ratio between estimated reach and sample spacing is more pessimistic than needed. Although it is difficult to determine by how much, the estimated reach may be substantially smaller than the actual reach of the signal manifold. In any event, it appears that the results of Sect. 6.2 (Figs. 6.10 and 6.11) are a reliable indication of the true topology of the signal manifold.

6.7 Open Questions

1. Persistent cohomology of general sheaves has not been studied extensively. We know that interval decomposition fails if coefficients aren't fields, but what else can be said?
2. We considered a linear chain of topological filters, one into the next. Most interesting signal processing chains have a branched structure. Is there a relevant interval decomposition (as in Carlsson, de Silva (2010)), and what is its interpretation? There is probably no such decomposition, though the study of the resulting zig-zag diagrams leads to *quiver theory*.
3. Eccentricity in the metric can cause problems for persistent homology. Depending on the interpretation, a nontrivial H_1 element arising points sampled from a thin cylinder (See Fig. 6.31) may or may not be significant. Is there a systematic way to assess the geometric sensitivity of these topological features?

4. It has been recently found that the space of persistence diagrams supports statistical constructs (Mileyko et al. 2011). Very little work has been done to explore the statistical properties of the resulting detectors. In order to use persistence diagrams as topological detectors in signal processing systems, it will be necessary to understand how they interface with traditional statistical signal models. Further, computing statistical parameters of interest such as a mean persistence diagram appears to be computationally infeasible. Are there approximations that requires less computation?
5. In order to verify the sampling criteria given in Theorem 6.4 in practice, it is necessary to estimate the reach of the manifold being reconstructed. How do errors in the reconstruction impact the estimate of its reach?

References

Amenta N, Choi S, Dey TK, Leekha N (2002) A simple algorithm for homeomorphic surface reconstruction. Int J Comput Geom Appl 12(1,2):125–141

Banyaga A, Hurtubise D (2004) Morse homology. Springer, Dordrecht

Carlsson G, de Silva V (2010) Zigzag persistence. Found Comput Math 10(4):367–405

Carrara W, Goodman R, Majewski R (1995) Spotlight synthetic aperture radar: signal processing algorithms. Artech House, Norwood

Chazal F, Lieutier A (2006) Topology guaranteeing manifold reconstruction using distance function to noisy data. In: Proceedings of the 22st symposium on computational geometry, p. 112–118

Cohen-Steiner D, Edelsbrunner H, Harer J (2007) Stability of persistence diagrams. Discrete Comput Geom 37(1):103–120

Derksen H, Weyman J (2005) Quiver representations. Not Am Math Assoc 52(2):200–206

Edelsbrunner H, Letscher D, Zomorodian A (2002) Topological persistence and simplification. Discrete Comput Geom 28:511–533

Federer H (1959) Curvature measures. Trans Amer Math Soc 93(3):418–491

Forman R (1998) Morse theory for cell complexes. Adv Math 134:90–145

Gabriel P (1972) Unzerlegbare darstellungen i. Manuscripta Math 6:71–103

Ghrist R (2008) Barcodes: the persistent topology of data. Bull Am Math Soc 45(1):61

Hatcher, A (2002) Algebraic topology. Cambridge University Press, Cambridge

Jakowatz C, Wahl D, Eichel P, Ghiglia D, Thompson P (1996) Spotlight-mode synthetic aperture radar: a signal processing approach, vol 101. Kluwer Academic Publishers, Boston

Kaczynski T, Mischaikow K, Mrozek M (2004) Computational homology. Springer, NY

Lee J (2003) Smooth manifolds. Springer, New York

Mileyko Y, Mukherjee S, Harer J (2011) Probability measures on the space of persistence diagrams. Inverse Prob 27(12):124007

Milnor J (1963) Morse theory. Princeton University Press, NJ

Mischaikow K, Nanda V (2013) Morse theory for filtrations and efficient computation of persistent homology. Discrete & Comput Geom 50(2):330–353

Niyogi P, Smale S, Weinberger S (2004) Finding the homology of submanifolds with high confidence from random samples

Patel A (2011) A continuous theory of persistence for mappings between manifolds. arXiv, preprint arXiv:1102.3395

Robinson M, Ghrist R (2012) Topological localization via signals of opportunity. IEEE Trans Signal Process 60(5):2362–2373

Sexton H, Vejdemo-Johansson M (2011) JPlex simplicial complex library. http://comptop.stanford.edu/programs/jplex/

Silva V de, Carlsson G (2004) Topological estimation using witness complexes. In: Alexa M, Rusinkiewicz S (eds) Eurographics symposium on point-based graphics

Silva V de, Morozov D, Vejdemo-Johansson M (2011) Persistent cohomology and circular coordinates. Discrete Comput Geom 45(4):737–759

Takens F (1981) Detecting strange attractors in turbulence. In: Rand DA, Young LS (eds) Dynamical systems and turbulence, p. 366–381. Springer, Berlin

Zomorodian A, Carlsson G (2005) Computing persistent homology. Discrete Comput Geom 33(2):249–274

Appendix A
Topological Spaces and Continuity

Topology gives sets a notion of convergent sequences. This is the appropriate setting for discussing iterative approximation. However, the minimal conditions for a space to be adequate for sequence convergence are too general for the applications discussed in this book. Most topologies that can be put on a set are in some sense pathological. This appendix serves to remind the reader of the definitions and key examples of commonly-used topological ideas, and is intended for quick reference rather than detailed study.

Definition A.1 A *topology* on a set X is a collection $\mathscr{T}$ of subsets of X that satisfy the following properties:

1. The empty set and X are both in $\mathscr{T}$,
2. Any union of elements in $\mathscr{T}$ is also in $\mathscr{T}$,
3. The intersection of any finite collection of elements in $\mathscr{T}$ is also in $\mathscr{T}$.

The pair $(X, \mathscr{T})$ is called a *topological space*. The elements in $\mathscr{T}$ are referred to as the *open sets* of $(X, \mathscr{T})$. Usually, the topology $\mathscr{T}$ is clear from context, in which case we will abuse notation and say "X is a topological space."

Example A.1 Given any set X, it can be endowed with the *discrete topology* $\mathscr{T}$ which consists of the set of all subsets of X. This topology is perhaps the least interesting topology, as it represents all points as being far from one another.

Example A.2 The *usual topology* for the real line $\mathbb{R}$ consists of all possible unions of intervals of the form $(a, b) \subset \mathbb{R}$ along with the empty set.

It is usually inefficient to specify a topology by listing all open sets. One easier way to construct a topology is by considering the *topology generated by a subset* $Y \subseteq X$, the intersection of all topologies containing Y. While theoretically convenient, the topology generated by a subset is perhaps worse from a practical point of view. Instead, we can construct a topology by taking unions of a smaller collection of sets, called its base.

M. Robinson, *Topological Signal Processing*,
Mathematical Engineering, DOI: 10.1007/978-3-642-36104-3,

Definition A.2 A collection $\mathscr{B}$ is called a *base* for a topology $\mathscr{T}$ if each $U \in \mathscr{T}$ is a union of elements of $\mathscr{B}$.

Proposition A.1 *If $\mathscr{B}$ is a base for $\mathscr{T}$, then the topology generated by $\mathscr{B}$ is $\mathscr{T}$.*

Most topologies that one encounters in applications are *Hausdorff*; every two points are contained in disjoint open neighborhoods.

Definition A.3 Suppose that $(X, \mathscr{T})$ is a topological space, and that $A \subset X$. We call the set $\mathscr{R} = \mathscr{T} \cap A$ the *subspace topology* on A, and call $(A, \mathscr{R})$ a topological subspace of $(X, \mathscr{T})$. Again, if the topologies are clear from context, we will usually abuse notation and say that "A is a subspace of X."

Definition A.4 Suppose that $(X, \mathscr{R})$ and $(Y, \mathscr{T})$ are topological spaces. A function $f : X \to Y$ is called *continuous* if $f^{-1}(U) \in \mathscr{R}$ for every $U \in \mathscr{T}$. For brevity, we usually call a continuous function a "map." A continuous bijection whose inverse is also continuous is called a *homeomorphism*.

Proposition A.2 *A homeomorphism places the open sets of two topological spaces in one-to-one correspondence.*

Homeomorphisms are therefore *topological equivalences*; anything which is true about a topological space $(X, \mathscr{T})$ is also true about any topological space homeomorphic to $(X, \mathscr{T})$.

The problem of determining if two topological spaces are homeomorphic is difficult in general; much of the theory of topology amounts to developing and classifying *invariants* that can discriminate between pairs of non-homeomorphic spaces. Cohomology (Chap. 4) is one such class of invariants. Dimension is another, weaker invariant, that is particularly effective on disks.

Proposition A.3 *(Invariance of dimension) An n-disk is homeomorphic to an m-disk if and only if $n = m$.*

Exercise A.1 Construct an explicit homeomorphism from the unit circle $\{(x, y) \in \mathbb{R}^2 : x^2 + y^2 = 1\}$ and the square $\{(x, y) \in \mathbb{R}^2 : \max\{|x|, |y|\} = 1\}$.

Definition A.5 Suppose that $f : X \to Y$ is an injective map between two topological spaces. If f is a homeomorphism between X and the subspace $f(X) \subseteq Y$, then f is called an *embedding*.

Example A.3 If $X \subseteq Y$ is a subspace, then the *inclusion map* $i : X \hookrightarrow Y$, which takes $x \mapsto x$ is an embedding.

Example A.4 A less trivial embedding is the map which takes the unit disk in $\mathbb{R}^2$ to the upper hemisphere in $\mathbb{R}^3$, given by $(x, y) \mapsto (x, y, \sqrt{1 - x^2 - y^2})$.

Definition A.6 A topological space $(X, \mathscr{T})$ is called *path connected* if for every pair of points $x, y \in X$, there exists a map from the closed unit interval $p : [0, 1] \to X$ for which $p(0) = x$ and $p(1) = y$.

Exercise A.2 Path connectedness can be unintuitive if the topological space is finite. Consider the topological space containing two points $X = \{0, 1\}$ with the following topology

$$\mathscr{T} = \{\emptyset, \{1\}, \{0, 1\}\}$$

Show that $(X, \mathscr{T})$ is path connected.

Definition A.7 In a topological space X, a subset A is called *closed* if its complement $X \backslash A$ is open.

Definition A.8 The *closure* of a subspace $Y \subseteq X$ is given by the set of points y for which $y \in U$ implies $U \cap Y$ is nonempty for every open U. The *interior* of a subspace $Y \subseteq X$ is the union of every open subset U of Y. The *frontier* or *boundary* of a subspace is given by the intersection of its closure and the complement of its interior.

Proposition A.4 *The closure of a subset $Y \subseteq X$ of a topological space is closed, and is the smallest closed set which contains Y.*

Definition A.9 Suppose that $(X, \mathscr{T})$ is a topological space and that there is an equivalence relation $\sim$ on the elements of X. Let $X/\sim$ be the set of equivalence classes of X under the relation $\sim$. For each element $x \in X$, we denote its equivalence class in $X/\sim$ by $[x]$. We define the *quotient topology* on $X/\sim$ to be

$$\mathscr{Q} = \{U \subseteq X/\sim : \bigcup_{[x]\in U} x \in \mathscr{T}\}.$$

That is, each element of $\mathscr{Q}$ is set of equivalence classes for which a union of their representatives is an open set in $\mathscr{T}$.

Exercise A.3 Graphs can be given topologies by using a quotient construction. Suppose $G = (V, E)$ is a graph, by which we mean $E \subseteq V \times V$. Consider the disjoint union X of $|E|$ open intervals in $\mathbb{R}$, which we associate with the edges of G. Construct an equivalence relation $\sim$ between points in these open intervals so that $X/\sim$ has the same topology as the realization (Definition 2.11) of $G = V \cup E$, considered as an abstract simplicial complex.

Definition A.10 If $(X, \mathscr{T})$ and $(Y, \mathscr{R})$ are topological spaces, then their *product* is a topological space $(X \times Y, \mathscr{T}\mathscr{S})$ where $\mathscr{T}\mathscr{S}$ is the topology generated by sets of the form $U \times V \subseteq X \times Y$ for $U \in \mathscr{T}$ and $V \in \mathscr{R}$.

Definition A.11 A collection of subsets $\mathscr{R}$ are said to *cover* another set Y if $Y \subset \cup\mathscr{R}$. A topological space $(X, \mathscr{T})$ is called *compact* if every collection $\mathscr{R} \subset \mathscr{T}$ of open sets which cover X has a finite subset $\mathscr{R}$ which also covers X.

Compactness is important because it ensures convergence of portions of sequences.

Proposition A.5 *Every infinite sequence* $\{x_n\}$ *of points in a compact space has a subsequence that converges.*

Proposition A.6 *(the Heine-Borel theorem) Compact subspaces of Euclidean space consist precisely of those sets that are closed and bounded.*

Every set can be made compact by adding an additional point. This idea is important to generalize from CW complexes to cell complexes in Chap. 2.

Proposition A.7 *(Alexandroff one-point compactification) Suppose that* $(X, \mathscr{T})$ *is a topological space. Then the topological space* $(X', \mathscr{T}')$ *constructed by adding another point* $*$ *to* X*, and*

$$\mathscr{T}' = \mathscr{T} \cup \{U \cup \{*\} : U \in \mathscr{T} \text{ and } X \backslash U \text{ is compact}\}$$

is a compact space.

Appendix B
Topological Groups

In practical settings, measurements often carry a group or group-like structure that permits them to be related to one another.

Definition B.1 A *group* is a set G with a binary operation $m : G \times G \to G$, that satisfies the following axioms:

1. (Associativity) For any three $a, b, c \in G$, $m(a, m(b, c)) = m(m(a, b), c)$.
2. (Identity) There is an element $1 \in G$ for which $m(1, a) = m(a, 1) = a$ for all $a \in G$.
3. (Existence of inverses) For each element $a \in G$, there is an element $a^{-1} \in G$ for which $m(a, a^{-1}) = m(a^{-1}, a) = 1$.

Usually, $m(a, b)$ is written ab whenever no confusion can arise. If $m(a, b) = m(b, a)$ for all $a, b \in G$, we call G an *abelian group* and we usually write $m(a, b) = a + b$.

It is important that measurements be *robust*, which means that although a measurement may be in error, it has nearly the correct value. An abstract notion of "nearly" is encoded in the definition of topology. When we use a group to represent measurements, we will usually want a topology on it that is compatible with the group operation.

Definition B.2 A *topological group* G is both a group (G, m) and a topological space $(G, \mathscr{T})$ in which the binary operation m and the inverse are both continuous.

A few examples of topological groups are listed in Table B.1 and outlined below, but the reader can imagine many more.

Example B.1 Any group with the discrete topology[1] is a topological group. This effectively removes the continuity requirement on the definition, but encompasses finite fields under addition and multiplication (excluding the zero) and the integers under addition.

[1] The topology in which all subsets are considered to be open. See Appendix A.

M. Robinson, *Topological Signal Processing*,
Mathematical Engineering, DOI: 10.1007/978-3-642-36104-3,

Table B.1 Several examples of topological groups used for representing measurements

Measurement type	Set	Operation	Abelian
Length, temperature, or mass	$\mathbb{R}$	Addition	Yes
Position	$\mathbb{R}^3$	Vector addition	Yes
Discretized timeseries	$\mathbb{Z}$	Addition	Yes
Magnitude and phase of a sinusoid	$\mathbb{C}$	Addition	Yes
Rotation angles	$n \times n$ orthogonal matrices with unit determinant $SO(n)$	Matrix Multiplication	No

Example B.2 $\mathbb{R}^n$ with the usual vector addition forms the group of *translations*, since its identity is the zero vector, and the function which takes v to $-v$ is the inverse operation.

- Vector addition is continuous. The usual topology on $\mathbb{R}$ can be generated by taking unions of open intervals (a, b). Suppose that $\varepsilon > 0$ and $w = u + v \in \mathbb{R}$ is given. If $\delta = \varepsilon/2$ and $u', v' \in \mathbb{R}$ satisfy $|u - u'| < \delta$ and $|v - v'| < \delta$, then

$$\begin{aligned} |(u' + v') - (u + v)| &= |(u' - u) + (v' - v)| \\ &\leq |u' - u| + |v' - v| \\ &\leq 2\delta = \varepsilon, \end{aligned}$$

  Repeating this argument componentwise proves that vector addition is continuous in the usual topology on $\mathbb{R}^n$.
- Additive inversion is continuous. Merely observe that $|-u - (-u')| = |u' - u| = |u - u'| = \varepsilon$ for any $\varepsilon > 0$ and u' chosen to satisfy the last equality.

Example B.3 By the previous example, complex numbers $\mathbb{C}$ under addition are a topological group. Complex numbers without 0 under multiplication are also a group. Recall that the complex numbers have a magnitude function $z \mapsto |z|$. This function generates the usual topology on $\mathbb{C}$ by taking unions of open disks $\{z \in \mathbb{C} : |z - z_0| < r\}$.

- Multiplication in $\mathbb{C}$ is continuous. Suppose that $\varepsilon > 0$ and $z = uv \in \mathbb{C}$ are given. Then if $u', v' \in \mathbb{C}$ are chosen so that

$$|u - u'| < \min\left\{\varepsilon/2, \frac{\varepsilon}{2|v| + \varepsilon}\right\}$$

  and

$$|v - v'| < \min\left\{\varepsilon/2, \frac{\varepsilon}{2|u| + \varepsilon}\right\},$$

then

$$\begin{aligned}|uv - u'v'| &= |uv - u'v' - u'v + u'v| \\ &= |(uv - u'v) + (u'v - u'v')| \\ &= |(u - u')v + u'(v - v')| \\ &\leq |u - u'||v| + |u'||v - v'| \\ &\leq 2\varepsilon.\end{aligned}$$

- Multiplicative inversion is continuous because $1/r$ is continuous when $r \neq 0$.

In almost all signal processing settings, topological vector spaces suffice to represent measurements. About the only exception is the case of the group of rotations.

Exercise B.1 The group of *rotations* in $\mathbb{R}^2$ is given by the unit complex numbers. This is continuous by the previous example. However, the group of rotations in $\mathbb{R}^3$ is given by the set of orthogonal 3×3 matrices with unit determinant under matrix multiplication, and is a non-abelian group. Topologically, the group of rotations has the subspace topology coming from the space of all 3×3 matrices with the usual topology. (The space of all 3×3 matrices is homeomorphic to $\mathbb{R}^9$.) Prove that the multiplication and matrix inversion are continuous under this topology. (Hint: you can write out these operations explicitly in terms of the matrix elements.)

Index

M. Robinson, *Topological Signal Processing*,
Mathematical Engineering, DOI: 10.1007/978-3-642-36104-3,

D

E

GPSR Compliance
The European Union's (EU) General Product Safety Regulation (GPSR) is a set of rules that requires consumer products to be safe and our obligations to ensure this.

If you have any concerns about our products, you can contact us on

ProductSafety@springernature.com

In case Publisher is established outside the EU, the EU authorized representative is:

Springer Nature Customer Service Center GmbH
Europaplatz 3
69115 Heidelberg, Germany

www.ingramcontent.com/pod-product-compliance
Ingram Content Group UK Ltd.
Pitfield, Milton Keynes, MK11 3LW, UK
UKHW021828190726
13853UKWH00003B/1249
* 9 7 8 3 6 6 2 5 2 2 8 4 4 *